ORCHIDS OF EASTERN GHATS (INDIA)

About the Author

Dr. Kothareddy Prasad obtained his M.Sc. (2008) and Ph.D. (2013) degrees in Botany from Sri Krishnadevaraya University, Ananthapuram. He was worked as a Post Doctoral fellow (SERB-DST, under young scientist scheme) at Botanical Survey of India, Deccan Regional Centre (2014-2017). At present he is working as AJCB-PDF at Central National Herbarium, Botanical Survey of India, Kolkatta since June 2017. Dr. K. Prasad has published 30 research papers in national and international journals. He described one new genus and 15 species new to science.

Dr. S. Karuppusamy obtained his M.Sc. (1996) degree from Madurai Kamaraj University and Ph.D. (2002) from Gandhigram University, Tamil Nadu. He was a Post-Doctoral Fellow at Sri Krishnadevaraya University (2000-2004). At present he is working as Assistant Professor, Department of Botany, The Madura College, Madurai since 2008. He has published 120 research papers and five books. He is the recipient of Young Botanist Award from Indian Botanical Society (2006). He received Prestigious Prof. Y.S. Murthy Gold Medal of Indian Botanical Society in 2013. He is also a recipient of Rolla S. Rao award and Fellow of Association for Angiosperm Taxonomy (FIAT) from Indian Association for Angiosperm Taxonomy. He is the Principal Investigator of major research projects funded from UGC, New Delhi and SERB, New Delhi.

Prof. T. Pullaiah obtained his M. Sc. (1973) and Ph. D. (1976) degrees in Botany from Andhra University. He was a Post-doctoral Fellow at Moscow State University, Russia, during 1976-1978. He joined Sri Krishnadevaraya University as Lecturer in 1979 and became Professor in 1993. He held several positions in the University as Dean, Faculty of Life Sciences; Head, Department of Botany; Chairman, Board of Studies in Botany; Head, Department of Sericulture; Co-ordinator and Chairman, Board of Studies in Biotechnology; Principal, Sri Krishnadevaraya University College. He retired from active Service in 2011 and was selected by UGC as UGC-BSR Faculty Fellow in the Department of Botany, Sri Krishnadevaraya University. He has published 73 books, 330 research papers and 35 popular articles.

ORCHIDS OF EASTERN GHATS (INDIA)

Kothareddy Prasad

Central National Herbarium
Botanical Survey of India
P.O. Indian Botanic Garden
Howrah - 711 103

Subbiah Karuppusamy

Department of Botany
The Madura College
Madurai - 625 011 (T.N.)

T. Pullaiah

Department of Botany
Sri Krishnadevaraya University
Anantapur - 515 003 (A.P.)

SCIENTIFIC PUBLISHERS

Published by

SCIENTIFIC PUBLISHERS (INDIA)
5 A, New Pali Road
P.O. Box 91
Jodhpur - 342 001 INDIA
E-mail: info@scientificpub.com
Website: http://www.scientificpub.com

ISBN: 978-93-87893-89-4 (Hardbound)
978-93-87991-64-4 (E-ISBN)

Printed in India

PREFACE

Orchids, the "Jewels" of plant kingdom, attracted the attention of both specialists and amateurs because of their beauty and complexity. The family Orchidaceae, is the most diverse family and account for eight percent of all flowering plants and majority are found in the tropics and subtropics. Orchids are generally forest living and prefer to grow in humid tropical climate. They were exploited till the end of last century ruthlessly for cut flowers and this has lead to their inclusion in the Appendices I and II of CITES. In India, the family Orchidaceae is the one of the largest family and represented by about 1331 taxa and about 400 are endemic. India is phytogeographically divided into three regions namely, Peninsular India, Himalayan region and Andaman Nicobar Islands. Peninsular India is again divided three regions namely: The Western Ghats, Deccan Peninsula and The Eastern Ghats.

The Eastern Ghats or Purva Ghat, also known as Mahendra Parvatam, are a discontinuous hill range of mountains run parallel to the Bay of Bengal. The Eastern Ghats encompass different types of vegetation and support rich Orchid flora. In the Eastern Ghats orchids species are at present in great danger of depletion and erosion due natural threats like climate change (unusual-rains, floods, temperature, and soil erosion), natural competition between species (native species and invasive species), lack of pollination (effect on seed formation) and natural degeneration *etc.* while man made threats include forest encroachment, mining activity, developmental activities, annual forest fires, grazing, illegal trade *etc.* Having realized the said issues there is urgent need to document orchids in Eastern Ghats and publish in a book form. At present there is no comprehensive account for Orchid flora of the Eastern Ghats. Authors taking note of this lacuna attempted to provide complete information supported by colour photographs.

We thank the authorities of Central National Herbarium (CAL), Howrah; Madras Herbarium (MH), Coimbatore; Deccan Regional Centre (BSID), BSI, Hyderabad; Forest Research Institute Herbarium (DD), Dehra Dun; authorities of Herbarium, Department of Botany, Sri Krishnadevaraya University, Anantapur; Rapinat Herbarium, Tiruchirapalli; Regional Plant Resource Centre, Bhubaneswar for permission to consult the Herbarium and Library. We thank to Dr. P. Singh, Director, Botanical Survey of India, Kolkata for support. We are thankful to Dr. M. Sanjappa, Dr. P. Venu and Dr. P.V. Prasanna, Scientists of Botanical Survey of India for their constant encouragement. Dr. K. Prasad would like to express his gratitude to the Director of Botanical Survey of India and SERB-DST, New Delhi for the Post Doctoral Fellowship.

We hope this can be source book for field identification and taxonomic research. Readers are requested to bring out deficiencies and mistakes, if any, so that they can be rectified during further editions.

K. Prasad
S. Karuppuamy
T. Pullaiah
E-mail: pullaiah.thammineni@gmail.com

02.07.2018

CONTENTS

CHAPTER 1

INTRODUCTION

EASTERN GHATS

The Eastern Ghats are generally considered as the hill ranges lying on the eastern side of the Deccan Plateau of Peninsular India. They are a chain of broken hill ranges in south-east India. The Eastern Ghats are located between 11°30' 22° N latitude and 76°50' – 86°30' E longitude in a north-east to south-west strike. The Ghats cover an area of about 75,000 sq.km. with an average width of 200 km in the north and 100 km in the south. They extend over a length of 1750 km between the rivers Mahanadi and Vaigai along the East coast. The geographers consider the Similipal massif lying to the north-west of the Khondmal hills in the Phulbani district of Odisha as the northern starting point of the Eastern Ghats. Towards the south, the Eastern Ghats run in a westward direction meeting the Western Ghats in the Nilgiris of Tamil Nadu.

Eastern Ghats are not a continuous range of mountains like the Western Ghats but an assemblage of a series of broken hills because the great rivers Mahanandi, Godavari, Krishna, Cauvery and Gundlakamma cut across them. Eastern Ghats are spread over five States of India, namely, Andhra Pradesh, Karnataka, Odisha, Tamil Nadu and Telangana (Fig. 1). Details of the important different hill ranges of Eastern Ghats are given below.

The Similipal (Similipahar) forests lying between 20°17' – 22°10' N and 87°57' – 86°47' E are spread over 2,750 sq. km. in Mayurbhanj district of Odisha. Mahendragiri hill ranges lying between 18°50' – 19°05' N and 84°10' – 84°30' E are physiographically bounded by Ramgiri and Udayagiri hills in the north, Gandahati Narayanpur in the south, Singaraja Parbat in the east and Devagiri hills in the west. The important mountain peaks of Mahendragiri range are Singaraja Parbat (Northwest of Mahendragiri hill), 1516 m above msl, Mahendragiri 1501 m, Devagiri 1382 m, Badadeba 1363 m, Sarango hill 1150 m and Gumma hill 1094 m.

The hill ranges in northern Andhra Pradesh are characterized by high rainfall, semi-deciduous forests in the lower, and evergreen forests in the higher ranges. Elevations at some places reach more than 1615 m. The Eastern Ghats of

with steep rocksides, a circular base and a plateau on top, covering an area of about 72 sq. km. and reaching a height of about 975 m. The Javadi hills form an interesting range of hills in North Arcot district and lies between 12°15' – 12°40' N and 78°2' – 79°10' E. The hills are running approximately north-east and south-west towards the eastern portion of Alagayam and Thirupattur. It attains maximum length of about 60 km and width of about 25 km. An interesting feature at the top of these hills is the place of plateaus extending for ten kilometers. The average height above mean sea level is 1053 m; the highest point is Kumbukudi (1273 m). The Burgur hills of Dharmapuri district lies in the north-eastern side of Erode district, between 11°28'–12° N and at 76°47' E and has an area of 64,729.6 ha. Elevation of the hills ranges between 975 and 1300 m.

The Kalrayan hills (11°38' – 12°01' N and 78°37' – 78°51' E), 25.76 km (north-south) and 37 km (east-west) long are located in Sankarapuram taluk of Villupuram district. The southern slopes are in the Salem and Dharmapuri districts while the major northern parts are in the Villupuram district. The Attur Kalrayas are divided into two by the valley running eastwards from Thumbal to Papinaickkanpatti, the northern part being Chinnakalrayans and the southern one the Periakalarayans. The total area occupied by these hills is 1,095 sq. km. and the altitude ranges between 300 and 1,000 m above m.s.l. The Chitteri hills are located towards northeast of Salem between 78°15' – 78°45' E and 11°44' -12°08' N and cover an area of 654.52 sq. km. The elevation is between 240 m and 1,270 m. The Chitteri hills occur in the Dharmapuri district, and form a compact block consisting of several hill ranges and contain tangled ridges and ravines running in the north-east and south-west directions enclosing many narrow valleys like Kallar, Varattar, Kambalai and Anaimaduru. They form the westward extension of the Kalrayan ranges. The Aranuthumalais (Salem district) are a south-western extension of the Chitteri hills.

The Aalagar hill comes under Alagar Koil range in Dindigul forest division. It is about 22 km east of Madurai city and it lies at 78°5' 30" E and 9°55' 10" N. It has a total area of 68.11 sq. km. The hill is about 16 km long with three valleys such as Periya aruvi valley, Silambar valley and Bison valley. The main watershed is Vaigainatham. The highest peak is Thalaianaiparai and is about 879 m above sea level.

The Sirumalai/Chinnamalai group of hills lies in four districts of Tamil Nadu: Tiruchirapalli, Dindigul, Karur and Sivagangai. The Sirumalai lies in Dindigul district and it is 16 km away from Dindigul. It's total area is 25.26 sq. km. It is situated between 77°33' – 78°15' E and 10°00' – 10°39' N. The Piranmalai hill is located in Sivagangai district and it covers an area of 90.23 sq. km. The Semmalai hills fall in Trichy and Dindigul districts and reach a maximum elevation of 1031 m. It forms one of the largest conglomerations of hillocks and dominate the Sirumalai group of hills.

The Melagiris are in the Dharmapuri district. It has the Melagiri peak (1225 m). These hills are enclosed between the rivers Cauvery and Chinnar. The Tamil Nadu Melagiris occupy about 2,910 sq. km. and form the south-eastern extension of the Karnataka tableland (parts of BR hills). The imposing hill ranges culminate in the Guthirayan peak (1,400 m) and the Sidhapuram peak (1,250 m) respectively. The Guthirayan hills can be approached from Palacode-Kodagarai shed. There is a shola with the Guthieri (tank) in the centre with dense forests.

Geology and Soils

The principal rock formations in the Eastern Ghats are charnokites, granite gneiss, khondalites, metamorphic gneisses and quarzites. Major soil types found in the Eastern Ghats are black soils, red soils and alluvial soils. The black soils can be seen in the Nallamalais, Veligonda hills, Seshachalam hills. Red soils are common throughout Eastern Ghats and occur widely in places such as Arumakonda, Chandragiri, Mahendragiri and Rampa hills. The alluvial soils are found in the coastal plains. In high altitudes of Kolli hills and Shervarayan hills considerable amount of humus is seen depending on the vegetation but basal rock soil becomes yellow with rich of iron and other minerals.

River systems

A large number of major and minor rivers flow through the Eastern Ghats. Except for the rivers Mahanadi and Godavari, all are seasonal. The major rivers are the Mahanadi, Godavari, Krishna and Cauvery. The minor rivers are Bahuda, Ryushikulya, Vamsadhara, Nagavali, Indravathi, Sileru, Sabari, Gundlakamma, Paleru, Sagileru, Pennar, etc. As the general terrain of the Eastern Ghats is inclined towards the west-east direction, all the rivers terminate in the Bay of Bengal.

Forests and Vegetation types

As per Champion and Seth (1968) classification of nine types of forest types are met within the Eastern Ghats as given below.

1. Evergreen Forests

This category of forests is seen to a very restricted extent, often along streams, only in a very few ‘sholas’/valleys in Similipal, Jeypore, Dhenkenal and Koraput areas of Orissa, in pockets of Lammasingi, Tajangi, Sapparla, Jerrela, Dharakonda, G. Madugula, Nallamalais and Seshachalam highlands in Andhra Pradesh, in the high altitudes of Shevaroy hills, Kolli hills, Pachamalai and Melagiri hills, all in Tamil Nadu and in the Bedguli ranges of Biligiri Rangana hills in Karnataka.

The common trees in the evergreen forests of Eastern Ghats are *Cinnamomum zeylanicum, Elaeocarpus serratus, Ixora notoniana, Meliosma microcarpa, Callicarpa tomentosa, Psydrax dicoccos, Symplocos cochinchinensis* var. *laurina, Schefflera capitata, Toona ciliata, Nothopegia heyneana, Neolitsea scrobiculata, Memecylon umbellatum, Embelia basaal,*

Breynia retusa, etc. The trees are with heavily moss and lichen-laden branches, harboring epiphytic orchids like *Bulbophyllum* spp., *Coelogyne* spp., *Conchidium* spp., *Dendrobium* spp., *Diploprora championii, Diplocentrum recurvum, Eria* spp., *Gastrochilus acaulis, Liparis* spp., *Oberonia* spp., *Pinalia polystachya, Taeniophyllum alwisii* etc. Occasionally some terrestrial orchids like *Acanthephippium sylhetense, Anoectochilus elatus, Calanthe* spp., *Chrysoglossum ornatum, Disperis zeylanica, Zeuxine* spp. *etc.* occur growing along with a few terrestrial ferns.

2. Tropical Semi-evergreen Forests

These are moist deciduous forests mixed with a greater proportion of evergreen and deciduous elements, giving the forests a semi-evergreen appearance. This type of forest is prevalent in moist valleys and on hills at about 800 m. Similipal forests of Mayurbhanj district, Atai and Banguri forests of Keonjhar district, some parts of Puri district, parts of Ganjam, Gajapati, Koraput, Malkangiri districts of Odisha, Sapparla, Dharakonda, Galikonda, Thanjavanam, Minumuluru, some areas near Anantagiri in Visakhapatnam district, Nulakamaddi and Maredumilli areas in East Godavari, Nallamalis and Seshachalam hills (all in the state of Andhra Pradesh), some high altitude areas of Kalrayan, Kolli hills, Shevaroy hills, Chitteri hills, Pacchamalais, Guthirayan hills, Sirumalai, Alagar hills etc. in Tamil Nadu and some parts of BR hills of Karnataka have semi-evergreen forest.

Trees such as *Magnolia champaca, Mangifera indica, Artocarpus lakoocha, Dillenia pentagyna, Firmiana colorata, Bridelia tomentosa, Xylia xylocarpa,* etc. form the top canopy, whereas small trees, like *Mesua ferrea, Polyalthia cerasoides, Macaranga peltata, Pittosporum nepalense, Phoebe lanceolata, Murraya koenigii* etc. form the middle storey.

Epiphytic orchids like, *Acampe* spp., *Aerides* spp., *Bulbophyllum* spp., *Chiloschista parishii, Cymbidium* spp., *Dendrobium* spp., *Oberonia* spp., *Phalaenopsis deliciosa, Sirhookera latifolia etc.* and terrestrial orchids like, *Acanthephippium bicolour, Cheirostylis* spp., *Crepidium* spp., *Dienia ophrydis, Liparis* spp., *Zeuxine* spp. *etc.* grow in these forests.

3. Tropical Moist Deciduous Forests

This can be sub-divided for convenience into

(A) Northern Tropical Moist Deciduous forests (Sal forests)

(B) Southern Tropical Moist Deciduous forests (non-Sal forests)

(C) Southern Tropical Moist Deciduous Riverian forests.

(A) Northern Tropical Moist Deciduous Forests (Sal forests)

This type of forest is found in parts of the districts of Kalahandi, Pulbani, Ganjam, Koraput (in Odisha), north-western parts of Srikakulam and North-eastern border area of Vizianagaram districts. These two Sal pockets in Andhra Pradesh represent the interesting southern limit of Sal extending from Madhya Pradesh through Koraput district. In the Sal forests *Shorea robusta* predominates and is associated with *Syzygium cumini, Xylia xylocarpa, Haldina cordifoliaa, Terminalia tomentosa, Pterocarpus marsupium, Anogeissus latifolia, Albizia procera, Madhuca longifolia* etc. forming the top canopy whereas the middle canopy is formed by trees like *Cleistanthus collinus, Buchanania lanzan, Dillenia pentagyna, Diospyros melanoxylon, Mallotus philippensis, Careya arborea, Syzygium articulatum* etc. It supports orchids like *Luisia* spp., *Habenaria* spp., *Peristylus* spp. etc.

(B) Southern Tropical Moist Deciduous Forests (non-Sal forests)

These forests are found in parts of Madgol hills, Gudem and Rampa Agency, Vizianagaram, East Godavari, West Godavari, Nallamalais and Seshachalam hills (all in Andhra Pradesh) and in parts of Tamil Nadu Eastern Ghats. *Tectona grandis* (in some parts only), *Terminalia alata, T. chebula, Xylia xylocarpa, Anogeissus latifolia, A. acuminata, Dillenia pentagyna, Haldina cordifolia, Sterculia urens, Bombax ceiba, Mitragyna parvifolia, Pterocarpus marsupium, Lagerstroemia parviflora, Schleichera oleosa, Mangifera indica, Dalbergia latifolia, Albizia odoratissima,* etc. form the top storey.

It supports epiphytic orchids like *Aerides* spp., *Cleisostoma* spp., *Cottonia peduncularis, Cymbidium* spp., *Luisia* spp., *Papilionanthe* spp., *Pelatantheria insectifera, Pholidota* spp., *Pomatocalpa* spp., *Rhynchostylis retusa, Taprobanea spathulata etc.* and terrestrial/saprophytic orchids like, *Aphyllorchis montana, Crepidium* spp., *Didymoplexis pallens, Dienia ophrydis, Eulophia* spp., *Geodorum* spp. *Habenaria* spp., *Liparis* spp., *Malaxis versicolor, Nervilia* spp., *Odisha cleistantha, Pecteilis* spp., *Peristylus* spp. *etc.*

(C) Southern Tropical Moist Deciduous Riverine Forests

This category of forests is found along the banks or rivers like Godavari, Krishna etc. and of major streams. These forests occupy only very limited areas. The most common trees in these forests are *Terminalia arjuna, Anogeissus acuminata, Tamarindus indica, Barringtonia acutangula etc.*

Few orchids like *Goodyera* spp., *Tainia penangiana, Tropidia* spp., *Zeuxine* spp. *etc.* grow in these forests.

4. Southern Tropical Dry Deciduous Forests

This climax type represents typical deciduous forest growing in larger areas along the Northern, Middle and Southern Eastern Ghats with distinct specific tree species growing predominantly. Thus, this type of forest is found in all the five States in which Eastern Ghats occur: Andhra Pradesh, Karnataka, Odhisha, Tamil Nadu and Telangana. Sal occurs in some forests of Srikakulam district and north of it, whereas teak occurs in some of the southern forests. The main components are sal or teak, *Terminalia alata, T. chebula, Pterocarpus marsupium, Anogeissus latifolia, Madhuca longifolia, Chloroxylon swietenia, Cassia fistula, Cleistanthus collinus, Bridelia retusa, Sterculia urens* etc. with shrubby layer of *Holarrhena pubescens, Grewia hirsuta, Helicteres isora, Catunaregum spinosa etc.*

The Red Sanders forests (*Pterocarpus santalinus*) are a pride of the Eastern Ghats thriving on the hill slopes and plateau of Seshachalam and Kambakkam hill ranges and southern portions of Nallamalais.

This type of habitat supports very few orchids like *Vanda* spp., *Eulophia* spp., *Habenaria* spp. *etc.*

5. Northern Mixed Dry Deciduous Forests

These forests are prevalent in Sukinda-Rebna, Keonjhargarh area, Migirdahuling area in Mayurbhanj district, parts of Angul division, parts of Kalahandi, Ganjam and Koraput districts. Sal may or may not be present, but normally *Boswellia serrata* is present, *Terminalia alata, Anogeissuss latifolia, Haldina cordifolia, Pterocarpus marsupium, Hymenodictyon excelsum, Bombax ceiba, Sterculia urens, Bridelia retusa etc.* form the top canopy. This type of habitat supports very few orchid species namely, *Eulophia* spp., *Habenaria* spp. *etc.*

6. Dry Savannah Forests

These forests, formed mostly as a result of intense biotic interference, are scattered throughout the area. Stunted tree species of *Phyllanthus emblica, Terminalia chebula, Pterocarpus marsupium* etc., with *Phoenix humilis* are common.

7. Tropical Thorn (Scrub) Forests

These forests are considered to be a result of intensive biotic interference. Such forests are seen in parts of Ganjam district, some drier areas of Andhra Pradesh, Tamil Nadu and Karnataka. *Albizia amara, Acacia chundra, Cassia fistula, Anogeissus latifolia, Chloroxylon swietenia, Catunaregum spinosa, Euphorbia tirucalli, E. antiquorum* are the common trees with shrubs such as *Dodonaea viscosa, Cassia auriculata, Dichrostachys cinerea, Capparis brevispina, Maytenus emarginatus, Carissa spinarum* etc. This type of forest supports few orchids, namely, *Eulophia epidendraea, E. graminea* and *Vanilla* spp.

8. Tropical Dry Evergreen Forest

This type of forest is seen in South Kadapa and Mamandur valley in Chittoor district in Andhra Pradesh, hills near Chennai, Chenglepet, and up to

Ramnad in Tamil Nadu. The tree species are *Manilkara hexandra, Syzygium cumini, Albizia amara, A. lebbeck, Strychnos nux-vomica, Sapindus emarginatus, Drypetes sepiaria, Pterospermum canescens* etc, while the middle layer comprises of *Diospyros ferrea, Memecylon umbellatum, Garcinia spicata, Cordia dichotoma, Flacourtia indica etc.*

FAMILY ORCHIDACEAE

Orchids form a unique group of plants where diversity is at its zenith. The enigma of these wonders has always been successful in creating inquisitiveness amongst both the educated and the commoners. Orchids have been gifted with numerous creative qualities of varying size, shape, colors and mimicry in flowers. Among all plant groups, they constitute a strange, yet interesting family and in almost all aspects Orchidaceae stands apart from the rest of the plant families, maintaining a fascinating individuality, all its own. The various morphological, physiological and genetic peculiarities are displayed by this group of plants have stimulated fields of research and varied that orchidology today has developed into one of the most dynamic branches of botany. The most intriguing feature of the orchid flowers is their outstanding resemblances to various forms of animal (seems to imitate) namely, bee orchid, butterfly orchid, moth orchid, dove orchid, donkey orchid, duck orchid, monkey orchid, lizard orchid, scorpion orchid, spider orchid, tiger orchid, man orchid and soldier orchids etc. Morphologically this group in very distinctive and are typically epiphytic/terrestrial/saprophytic in habit, bilateral symmetry, resupinate/non-resupinate flowers, a labellum/lip (a modified petal), a gynostegium (a columnar structure formed due to adnation of androecium and gynoecium), pollinia, minutest and non-endospermic seeds and their germination needs association with the mycorrhizal fungi. The family, Orchidaceae represents a peak in the evolution of monocots and one of the most successful families of flowering plants. The family members have wide distribution and the innumerable number of species are spread all over the world but majority of them in tropics and subtropics.

They were earlier used for their therapeutic properties to cure various ailments. Besides medicinal properties, perfumery, cosmetology, horticulture, food and ornamentation aspects make the orchid industry as one of the highest remunerative commercialization around the globe in cut flowers. India has a very rich repository of orchids *i.e.* about 1350 orchid species in about 185 genera. The North Eastern Himalayan, the Peninsular and the Andaman and Nicobar regions constitute the orchid-rich belts in the country. Due to lack of awareness and non-availability of proper cultivation procedures, commercial cultivation of orchids is still in its infancy. Conventional and biotech methods of improvement have added in the growth of economy of countries like Thailand, Malaysia, and Singapore etc. Appropriate management, improvement and sustainable

Rheede (1678-1693) who described 17 orchid taxa from Malabar. Roxburgh (1832) described 57 species of orchids in 8 genera. Lindley (1840) on the basis of collections made by Nathaniel Wallich and the latter predecessors described all the known genera and species of Indian orchids. Wight (1845-1853) Icons contain invaluable information regarding Indian orchids. Griffith (1851) described 147 species of orchids in 49 genera, and provided illustrated descriptions which covered many orchids of Khasi hills. Lindley (1857) described 71 species of Indian orchids in 21 genera. Reichenbach (1862) acquired many Indian orchids from various sources and described. Beddome (1874) provided an excellent drawing and a very accurate description of lady slipper orchid, *Cypripedium druryi* Bedd.

Indian Orchidology reached its pinnacle with the celebrated work of Hooker (1888, 1890). Hooker made very accurate descriptions of some 1200 species of orchids belonging to 113 genera in the British India. Even today, this work is considered to be the most exciting source on the subject. Hooker (1895) brought out an illustrated account of Indian orchids. It contains descriptions and illustrations of 100 species. King and Pantling (1898) described many new orchid species from Sikkim and Sikkim Himalaya. The works of Hooker, King and Pantling and Duthie are considered as the monumental treatments on Indian orchids.

After Independence with the reorganization of the Botanical Survey of India, exploration of orchid wealth in India has increased. Pradhan (1976, 1979) enumerated 810 orchid species under 130 genera and also provided information on orchid cultivation and major orchid habitats in India. Bose and Bhattacharjee (1980) reported 834 orchid species under 127 genera and outlined the cultural procedures for many species. Abraham and Vatsala (1981) reported about 150 orchids species, with line drawings of South India and provided cytological aspects, cultivation and classification. Jain and Mehrotra (1984) gave a preliminary inventory of Orchidaceae in India. An account on endemic orchids of South India was given by Kumar and Sasidharan (1986). Karthikeyan *et al.* (1989) provided a checklist of 1153 orchid taxa from India. Satish Kumar and Manilal (1994) published catalogue of Indian orchids comprising 1141 species belonging to 166 genera. Misra (2007) made a critical study on Indian orchids and recorded about 1331 orchid taxa in 184 genera. An account on Ladys' slipper orchids of India was given by Chowdhery (2015). Jalal and Jayanthi (2012) reviewed the endemic orchids of Peninsular India and reported that 130 orchid species are endemic to Peninsular India and 22 species endemic to the Eastern Ghats. The orchid-rich regions in India are the North-Eastern region, particularly the Eastern Himalaya, Meghalaya, Mizo or Lushai hills and the Naga hills; Sikkim in particular is rich with orchids; North-Western Himalayan region and also Western Ghats, the detached Pulney, Nilgiri and Biligirirangan hills (Rao, 1979). Sarkar (1995) gave the details of rare, endangered and endemic orchids in India. Satish Kumar and Manilal (1992) listed 630 species of Epiphytic orchids in India spread over 85 genera. Hegde (1997) reported that of the 1100 species of orchids occurring in India 314 are endemic, 215 are endangered and 14 are nearly extinct.

Duthie (1906) made a comprehensive study of the orchids of North-Western Himalayas an account of 173 species belonging to 45 genera. Bruhl (1926) wrote a guide to the orchids of Sikkim Himalayas wherein 461 species belonging to 92 genera were described. Raizada *et al.* (1981) reported about 89 species in 35 genera from Mussoorie. Seidenfaden and Arora (1982) enumerated 250 species of orchids of the North-Western Himalayas. Deva and Naithani (1986) reported about 239 orchids of North-West Himalayas.

Many orchidologists studied the orchids of North-East India. Ghatak and Devi (1986) reported 207 taxa under 67 genera from Manipur while Sathish Kumar and Suresh Kumar (2005) reported 280 species under 86 genera from the same state of Manipur. Kataki (1986) reported 324 species from Meghalaya. Hynniewta *et al.* (2000) reported 241 species spread over 63 genera from Nagaland. Chowdhery (1998) reported 545 species spread over 125 genera from Arunachal Pradesh. Rao (2007) listed a total of 900 orchid species belonging to 165 genera in North-East India. According to him Arunachal Pradesh has maximum diversity with 568 species (141 genera) followed by Sikkim with 532 species (138 genera), Meghalaya 361 species (107 genera), Nagaland 348 species (98 genera), Manipur 296 species (87 genera), Mizoram 237 species (81 genera), Assam 145 species (59 genera) and Tripura 35 species (27 genera).

Cooke (1906) reported 74 orchids under 31 genera from the Presidency of Bombay; Sharma *et al.* (1984) enumerated 173 species of orchids in 51 genera from the Karnataka state. Lakshminarasimhan (1996) recorded 110 orchid species (34 genera) in Maharashtra. Pande *et al.* (2010) gave an account of wild orchids of North-Western Ghats. Rao (1914) enumerated 171 species of orchids from Travancore and also discussed medicinal values of some orchids. Sathish Kumar and Manilal (2004) enumerated 256 orchid taxa belonging to 79 genera in Kerala. Sasidharan (2004) listed a total of 264 orchid taxa in Kerala. Joseph (1987) enumerated 116 orchid species and 49 genera of Nilgiris. Henry *et al.* (1989) listed 200 orchid taxa from Tamil Nadu. Seidenfaden (1999) described 81 species belonging to 39 genera encountered in Palni Hills, Tamil Nadu and clarified some of the taxonomic problems of South Indian orchids.

Many revisionary works on various genera have been published. These include Das and Jain (1980) on *Coelogyne*, Ansari and Balakrishnan (1990) on *Oberonia*, De and Hajra (2001, 2004) on *Tropidia* and *Vanda* and Phukan (2011) on *Cleisostoma*. A Pictorial Compendium on Wild Orchids in Karnataka was recently brought out by Ananda Rao and Sridhar (2007). Vij *et al.* (2013) brought out a book on orchids of Himachal Pradesh. A Pictorial Guide on Wild Orchids of Meghalaya was given by Rao and Singh (2015).

Studies on Orchidaceae of Eastern Ghats

There is no comprehensive work on Orchids of Eastern Ghats till date. The botanical history of the Eastern Ghats dates back to 1795-1819 when William Roxburgh, who surveyed plants of the Coast of Coromandel, published *Plants of the Coromandel Coast* in 3 volumes. In this work he described 9 orchid species under 4 genera, namely, *Orchis plantaginea* [**Habenaria roxburghii** Nicolson], *Limodorum virens* [**Eulophia epidendraea** (J. Koenig *ex* Retz.) C.E.C.Fisch.], *L. recurvum* [**Geodorum recurvum** (Roxb.) Alston], *L. nutans* [**Geodorum densiflorum** (Lam.) Schltr.], *Epidendrum tessellatum* [**Vanda tessellata** (Roxb.) Hook. *ex* G. Don], *E. praemorsum* [**Acampe praemorsa** (Roxb.) Blatt. & McCann], *E. pendulum* [**Cymbidium aloifolium** (L.) Sw.] and *Aerides multiflorum.*

Fischer (1928) in Gamble's *Flora of Presidency of Madras* provided accurate description of 199 orchid species belonging to 60 genera and most of the species was distributed in Eastern Ghats Part. Reddy *et al.* (2005) listed a total of 190 orchid species belonging to 54 genera that are reported from Eastern Ghats and in this list presently many names are synonyms.

Seidenfaden (1983) described 73 species belonging to 36 genera in flora of Tamil Nadu Carnatic based on collections made by K.M. Matthew and others. He has clarified some of the taxonomic problems of South Indian orchids; reported two Sri Lankan endemic species for the first time from India, viz., *Gastrochilus acaulis* and *Taeniophyllum alwisii* and one new record *Anoectochilus elatus* for the Eastern Ghats. Henry *et al.* (1989) listed 200 orchid taxa from Tamil Nadu which includes Eastern Ghats part also. Ansari and Dwarakan (2002) reported 66 orchid species found in the Shevoroy and Kolli hills in south Eastern Ghats. Misra (2012) described and illustrated a new species, *Tropidia hegderaoi* from Kolli hills of Tamil Nadu. Kottaimuthu and Ponnuchamy (2012) reported *Bulbophyllum tremulum* Wight as a new record for Eastern Ghats. Karuppusamy and Ravichandran (2013) reported *Eria exilis* Hook.f. as a new record for Eastern Ghats while Panda and Misra (2013) reported *Phreatia elegans* Lindl. (Orchidaceae) as a new record for Eastern Ghats. Saravanan and Kaliamoorthy (2017) reported *Zeuxine chowdheryi* Av. Bhattacharjee & Sabap. and *Z. grandis* Seidenf. as new records for the Eastern Ghats.

Saxena and Brahmam (1995) recorded a total of 123 species belonging to 41 genera from Odisha state. Misra (2004) published a classical work on orchids of Odisha and recorded a total of 130 orchid species belonging to 58 genera; discussed about classification, ecology, distribution pattern and clarified many taxonomic problems. Misra (2012) described and illustrated a new species, *Zeuxine mooneyi* from Rebana forest of Odisha. Dash *et al.* (2015) reported *Cheirostylis parvifolia* Lindl. as a new record for Eastern Ghats from Mahendragiri hills.

Ramakrishna (1997) recorded 67 orchid species belonging to 34 genera from undivided Andhra Pradesh while Pullaiah and Karuppusamy (2008) added three more species. Savithramma (2005) gave an account of orchids of Talakona Sacred Grove in Seshachalam hills. Reddy *el al.* (2007) reported *Sirhookera latifolia* (Wight) Kuntze as a new generic record for Eastern Ghats of Andhra Pradesh. Misra *et al.* (2008) reported 83 species of orchids belonging to 40 genera from Andhra Pradesh and provided an updated nomenclature. V.S. Raju *et al.* (2008) provided a list of 77 species from Andhra Pradesh. Subba Rao and Kumari (2008) recorded 23 species belonging to 18 genera from Visakhapatnam district. Karuppusamy *et al.* (2009) reported *Habenaria commelinifolia* from Tirumala hills. Ravi Prasad Rao *et al.* (2009) listed 56 orchids of Eastern Ghats of Andhra Pradesh that are part of East Himalayan orchids. Sadasivaiah *et al.* (2009) reported *Habenaria panigrahiana* as a new distributional record for Andhra Pradesh. Prasad and Ravi Prasad Rao (2010) reported *Geodorum recurvum* as a new distribution record for Southern Eastern Ghats of India. Prasad *et al.* (2010) reported two species, *Liparis nervosa* and *L. paradoxa* as new distributional records for Andhra Pradesh and also provided a key for all *Liparis* species of Andhra Pradesh. Ravi Prasad Rao *et al.* (2010) reported the occurrence of *Eulophia flava* in Eastern Ghats as well as southern peninsular India. Sadasivaiah *et al.* (2010) relocated three rare orchids, *Eulophia graminea*, *E. ochreata* and *Habenaria barbata* in Andhra Pradesh after lapse of eight decades. Ravi Prasad Rao and Prasad (2011) reported *Geodorum laxiflorum* as a new distributional record for southern peninsular India along with detailed description and photos. Reddy *et al.* (2014) reported *Thunia alba* (Lindl.) Rchb.f. var. *bracteata* (Roxb.) N. Pearce & P.J. Cribb as a new generic record for Andhra Pradesh from Narsipatnam Forest Division of Visakhapatnam district. Prasad and Veeranjaneyulu (2016) reported 27 species of orchids under 14 genera from Seshachalam Biosphere Reserve.

Pullaiah (2015) in his *Flora of Telangana* reported 12 species of orchids belonging to 7 genera. This new state was earlier in the state of Andhra Pradesh. Parts of Eastern Ghats extend into this state in the districts of Mahabubnagar, Khammam and Nalgonda. Ramachandrachary and Wood (1981) described a new species of orchid *Habenaria ramayyana* from Telangana and Kumar *et al.* (2016) reduced this species as a heterotypic synonym of *Habenaria panigrahiana*.

ABBREVIATIONS

States

AP: Andhra Pradesh
OD: Odisha
TE: Telangana
TN: Tamil Nadu
KA: Karnataka

Districts

CTR: Chittoor
EG: East Godavari
KDP: Kadapa (formerly Cuddapah)
KNL: Kurnool
NLR: Nellore
PKM: Prakasam

SKLM: Srikakulam
VSKP: Visakhapatnam
KMM: Khammam
MBNR: Mahabubnagar
CPT: Chingleput
DGL: Dindigul
DMP: Dharmapuri
NK: Namakkal
SA: South Arcot
SLM: Salem
TRP: Tiruchirapally
TVM: Tiruvannamalai
VLR: Vellore
CRN: Chamarajanagar
BLW: Baleswar
DK: Dhenkanal
GJM: Ganjam
GJP: Gajapathi
KDM: Kandhamal
MB: Mayurbhanj
KH: Kalahandi
KP: Koraput
KJ: Kendujhar
RG: Rayagada

Plant Collectors

AD: A. Diraviadoss
AM: A. Mohan
AMR: A.Madhusudana Reddy
ANH: A.N.Henry
ASR: A.Srinivasa Rao
BR: B.Ravi Prasad Rao
BSN: B. Suryanarayana
BSS: B. Sadasivaiah
CAB: C.A.Barber
CECF: C.E.C.Fischer
CM: C. Manoharan
CR: C. Rajamani
CSR: C. Sudhakar Reddy
DAM: D.Ali Moulali
DDSR: D. Daniel Sunderaraj
DIA: D.I. Arockiasamy
DN: D.Narasimhan
DR: D. Rangachari
EV: E. Vajravelu
GP: G. Paramasivan
GVS: G.V. Subba Rao
HFM: H.F. Mooney
JLE: J.L. Ellis
JSG: J.S. Gamble
KCJ: K.C. Jacob
KM: K. Murugesan
KMM: K.M. Matthew
KMS: K.M. Sebastine
KNR: K.N. Rama Krishna
KP: K. Prasad
KRK: K. Ravikumar
KS: K. Sampathkumar
KTA: K.T. Angustine
KTM: K.T. Mathew
MBV: M.B. Viswanathan
MC: M. Charles
MM: M. Mohanan
MSR: M.S. Rama Swami
MV: M. Venkaiah
MVS: M.V. Suresh
NPBK: N.P. Balakrishnan
NR: N. Rajendran
NRI: N. Rani
NRR : N. Rama Rao
NV: N. Venugopal
PP: P. Perumal

PS: P. Santhan
RCS: R. Chandrasekaran
RVR: R.R. Venkata Raju
RVS: R. Vijayasankar
SA: S. Alamelu
SJB: S. John Britto
SKB: S. Khadar Basha
SKS: S. Karuppusamy
SKW: S.K. Wagh
SM: Sarat Misra
SP: S. Periyanayagam
SRSR: S. Rajasekhara Reddy
SS: S. Sudhakar
STR: S.T. Ramachandrachary
TP: T. Pullaiah
TRS: T. Ravishankar
TSJ: T.S. Jayaseelan
VAA: V. Alphonse Amalraj
VBH: V.B. Hosagoudar
VKN: V. Karunanidhi
VNS: V. Narayanaswamy

ANALYSIS OF ORCHIDS OF EASTERN GHATS

A total of 197 taxa belonging to 66 genera were considered in the present Eastern Ghats and this excludes those species which have doubtful occurrence in Eastern Ghats. Life form analysis of orchid taxa in the Eastern Ghats revealed the presence of four different types of the life forms namely: epiphytes or lithophytes, terrestrial, vines and saprophytes. Of the 197 orchid taxa, epiphytes and lithophytes dominate with 101 taxa belonging to 38 genera followed by terrestrial orchid species are 91 belonging to 25 genera, vines are 2 species belonging to *Vanilla* genus and saprophytes are 3 species belonging to 3 genera.

In epiphytic forms the largest genera are the *Dendrobium* with 19 species followed by *Bulbophyllum* with 12 species, *Oberonia* with 10 species, *Luisia* with 5 species, *Acampe* and *Aerides* (4 species), *Conchidium, Eria, Gastrochilus, Liparis* (3 species), *Cleisostoma, Coelogyne, Cymbidium, Papilionanthe, Pholidota, Pomatocalpa* and *Vanda* each with 2 species. The genera *Callostylis, Chiloschista, Cottonia, Diplocentrum, Diploprora, Micropera, Pelatantheria, Phalaenopsis, Phreatia, Pinalia, Polystachya, Porpax, Rhynchostylis, Schoenorchis, Seidenfadeniella, Sirhookera, Smitinandia, Staurochilus, Taeniophyllum, Taprobanea* and *Thunia* are represented in Eastern Ghats each with single species.

In the terrestrial forms the most dominated genera is *Habenaria* with 22 species followed by *Zeuxine* with 9 species, *Liparis* with 8 species, *Eulophia* with 7 species,, *Nervilia* with 6 species, *Peristylus* (5 species), *Geodorum* and *Goodyera* each with 4 species, *Crepidium* and *Tropidia* each with 3 species and *Acanthephippium, Calanthe, Cheirostylis, Malaxis* and *Pecteilis* each with 2 species. The genera belonging to *Anoectochilus, Chrysoglossum, Corymborkis, Dienia, Disperis, Odisha, Pachystoma, Satyrium, Spiranthes* and *Tainia* each with a single species. In Eastern Ghats only one vine genus, *Vanilla* is encountered with 2 species and the saprophytic genera namely, *Aphyllorchis*,

Didymoplexis and *Epipogium* each with single species. In India, five largest orchid genera (Misra, 2007) are *Dendrobium* with 116 species, followed by *Bulbophyllum* (100 species), *Habenaria* (83 species), *Oberonia* (65 species) and *Eria* with 60 species and interestingly in the Eastern Ghats the first two dominant genera of India also have shown their dominance in the study area: *Habenaria* with 22 species followed by *Dendrobium* (19 species), *Bulbophyllum* (12 species) and *Oberonia* with 10 species.

State-wise analysis

SN	Name of taxon	Habit	AP	KA	OD	TE	TN
1	*Acampe carinata* (Griff.) Panigrahi	E	+		+		
2	*Acampe ochracea* (Lindl.) Hochr.	E	+		+		
3	*Acampe praemorsa* (Roxb.) Blatt. & McCann	E	+	+	+		+
4	*Acampe rigida* (Buch.-Ham. *ex* Sm.) P.F. Hunt	E	+		+		
5	*Acanthephippium bicolor* Lindl.	T	+		+		+
6	*Acanthephippium sylhetense* Lindl.	T			+		
7	*Aerides maculosa* Lindl.	E	+		+		
8	*Aerides multiflora* Roxb.	E	+		+		
9	*Aerides odorata* Lour.	E	+		+		
10	*Aerides ringens* (Lindl.) C.E.C. Fisch.	E	+	+	+		
11	*Anoectochilus elatus* Lindl.	T					+
12	*Aphyllorchis montana* Rchb.f.	S	+				
13	*Bulbophyllum cariniflorum* Rchb.f.	E	+		+		
14	*Bulbophyllum crassipes* Hook.f.	E			+		
15	*Bulbophyllum fischeri* Seidenf.	E					+
16	*Bulbophyllum guttulatum* (Hook.f.) N.P. Balakr.	E			+		
17	*Bulbophyllum kaitiense* Rchb.f.	E	+				+
18	*Bulbophyllum macraei* (Lindl.) Rchb.f.	E			+		
19	*Bulbophyllum panigrahianum* S. Misra	E			+		
20	*Bulbophyllum polyrrhizum* Lindl.	E			+		
21	*Bulbophyllum sterile* (Lam.) Suresh	E	+				+
22	*Bulbophyllum tremulum* Wight	E					+

SN	Name of taxon	Habit	AP	KA	OD	TE	TN
23	*Bulbophyllum triste* Rchb.f.	E			+		
24	*Bulbophyllum umbellatum* Lindl.	E	+		+		
25	*Calanthe triplicata* (Willemet) Ames	T			+		+
26	*Calanthe sylvatica* (Thouars) Lindl.	T			+		
27	*Callostylis bambusifolia* (Lindl.) S.C. Chen & J.J. Wood	E	+		+		
28	*Cheirostylis flabellata* (A. Rich.) Wight	T					+
29	*Cheirostylis parvifolia* Lindl.	T			+		
30	*Chiloschista parishii* Seidenf.	E			+		
31	*Chrysoglossum ornatum* Blume	T					+
32	*Cleisostoma appendiculatum* (Lindl.) Benth. & Hook.f. *ex* B.D. Jacks.	E			+		
33	*Cleisostoma tenuifolium* (L.) Garay	E	+				
34	*Coelogyne breviscapa* Lindl.	E		+			+
35	*Coelogyne nervosa A.Rich.*	E		+			
36	*Conchidium exile* (Hook.f.) Ormerod	E					+
37	*Conchidium nanum* (A. Rich.) Brieger	E					+
38	*Conchidium reticosum* (Wight) Ormerod	E					+
39	*Corymborkis veratrifolia* (Reinw.) Blume	T	+				
40	*Cottonia peduncularis* (Lindl.) Rchb.f.	E	+		+		
41	*Crepidium acuminatum* (D. Don) Szlach.	T	+				+
42	*Crepidium mackinnonii* (Duthie) Szlach.	T			+		
43	*Crepidium purpureum* (Lindl.) Szlach.	T			+		
44	*Cymbidium aloifolium* (L.) Sw.	E	+		+		+
45	*Cymbidium bicolor* Lindl.	E			+		
46	*Dendrobium anamalayanum* Chandrab.	E					+
47	*Dendrobium aphyllum* (Roxb.) C.E.C. Fisch.	E	+	+	+		
48	*Dendrobium aqueum* Lindl.	E	+				+
49	*Dendrobium bicameratum* Lindl.	E			+		
50	*Dendrobium cathcartii* Hook.f.	E			+		
51	*Dendrobium crepidatum* Lindl. & Paxton	E			+		

SN	Name of taxon	Habit	AP	KA	OD	TE	TN
52	*Dendrobium formosum* Roxb. *ex* Lindl.	E			+		
53	*Dendrobium herbaceum* Lindl.	E	+		+		+
54	*Dendrobium heterocarpum* Wall. *ex* Lindl.	E		+			+
55	*Dendrobium macraei* Lindl.	E			+		
56	*Dendrobium macrostachyum* Lindl.	E	+	+	+	+	
57	*Dendrobium microbulbon* A. Rich.	E					+
58	*Dendrobium moschatum* (Buch.-Ham.) Sw.	E	+		+		
59	*Dendrobium nanum* Hook.f.	E					+
60	*Dendrobium nodosum* Dalzell	E					+
61	*Dendrobium peguanum* Lindl.	E			+		
62	*Dendrobium regium* Prain	E	+		+		
63	*Dendrobium transparens* Wall. *ex* Lindl.	E			+		
64	*Dendrobium wightii* A.D. Hawkes & A.H. Heller	E					+
65	*Didymoplexis pallens* Griff.	S	+		+		
66	*Dienia ophrydis* (J. Konig) Seidenf.	T	+		+		
67	*Diplocentrum recurvum* Lindl.	E					+
68	*Diploprora championii* (Lindl.) Hook.f.	E			+		
69	*Disperis zeylanica* Trimen	T					+
70	*Epipogium roseum* (D. Don) Lindl.	S					+
71	*Eria lasiopetala* (Willd.) Ormerod	E			+		
72	*Eria meghasaniensis* (S. Misra) S. Misra	E			+		
73	*Eria pauciflora* Wight	E		+			+
74	*Eulophia epidendraea* (J. Konig *ex* Retz.) C.E.C. Fisch.	T	+			+	+
75	*Eulophia explanata* Lindl.	T	+		+		
76	*Eulophia flava* (Lindl.) Hook.f.	T	+				
77	*Eulophia graminea* Lindl.	T	+		+	+	+
78	*Eulophia ochreata* Lindl.	T	+		+		
79	*Eulophia pulchra* (Thouars) Lindl.	T					+
80	*Eulophia spectabilis* (Dennst.) Suresh	T	+	+	+	+	

SN	Name of taxon	Habit	AP	KA	OD	TE	TN
81	*Gastrochilus acaulis* (Lindl.) Kuntze	E		+	+		+
82	*Gastrochilus calceolaris* (Buch.-Ham. *ex* Sm.) D. Don	E	+				+
83	*Gastrochilus inconspicuus* (Hook.f.) Kuntze	E			+		
84	*Geodorum attenuatum* Griff.	T			+		
85	*Geodorum densiflorum* (Lam.) Schltr.	T	+	+	+	+	+
86	*Geodorum laxiflorum* Griff.	T	+		+		
87	*Geodorum recurvum* (Roxb.) Alston	T	+		+	+	
88	*Goodyera fumata* Thwaites	T			+		
89	*Goodyera hispida* Lindl.	T			+		
90	*Goodyera procera* (Ker-Gawl.) Hook.	T	+		+		
91	*Goodyera thailandica* Seidenf.	T			+		
92	*Habenaria barbata* Wight *ex* Hook.f.	T	+		+		+
93	*Habenaria crassifolia* A. Rich.	T			+		
94	*Habenaria commelinifolia* (Roxb.) Wall. *ex* Lindl.	T	+		+		
95	*Habenaria crinifera* Lindl.	T					+
96	*Habenaria digitata* Lindl.	T	+		+		
97	*Habenaria diphylla* (Nimmo) Dalzell	T	+		+		
98	*Habenaria furcifera* Lindl.	T	+	+	+	+	
99	*Habenaria grandifloriformis* Blatt. & McCann	T			+		+
100	*Habenaria heyneana* Lindl.	T					+
101	*Habenaria hollandiana* Santapau	T	+				
102	*Habenaria longicorniculata* J. Graham	T	+		+		+
103	*Habenaria longicornu* Lindl.	T					+
104	*Habenaria marginata* Colebr.	T	+		+	+	
105	*Habenaria multicaudata* Sedgw.	T					+
106	*Habenaria ovalifolia* Wight	T	+				+
107	*Habenaria panigrahiana* S. Misra	T	+		+	+	+
108	*Habenaria plantaginea* Lindl.	T	+	+	+	+	+
109	*Habenaria rariflora* A. Rich.	T					+

SN	Name of taxon	Habit	AP	KA	OD	TE	TN
110	*Habenaria reniformis* (D. Don) Hook.f.	T			+		
111	*Habenaria roxburghii* Nicolson	T	+	+	+	+	+
112	*Habenaria stenopetala* Lindl.	T	+		+		
113	*Habenaria viridiflora* (Rottler *ex* Sw.) Lindl.	T					+
114	*Liparis atropurpurea Lindl.*	T					+
115	*Liparis biloba* Wight	T					+
116	*Liparis deflexa* Hook.f.	T	+				
117	*Liparis downii* Ridl.	T			+		
118	*Liparis elliptica* Wight	E	+		+		
119	*Liparis espeevijii* S. Misra	E			+		
120	*Liparis nervosa* (Thunb.) Lindl.	T	+		+		
121	*Liparis odorata* (Willd.) Lindl.	T	+		+	+	
122	*Liparis viridiflora* (Blume) Lindl.	E	+		+		+
123	*Liparis walkeriae* Graham	T					+
124	*Liparis wightiana* Thwaites	T					+
125	*Luisia abrahamii* Vatsala	E					+
126	*Luisia birchea* Blume	E					+
127	*Luisia brachystachys* (Lindl.) Blume	E			+		
128	*Luisia trichorrhiza* (Hook.) Blume	E	+		+	+	
129	*Luisia zeylanica* Lindl.	E	+	+	+	+	+
130	*Malaxis intermedia* (A. Rich.) Seidenf.	T		+			
131	*Malaxis versicolor* (Lindl.) Abeyw.	T	+		+		+
132	*Micropera pallida* (Roxb.) Lindl.	E			+		
133	*Nervilia aragoana* Gaudich.	T	+	+	+	+	
134	*Nervilia crociformis* (Zoll. & Moritzi) Seidenf.	T	+		+		+
135	*Nervilia falcata* (King & Pantl.) Schltr.	T			+		
136	*Nervilia gammieana* (Hook.f.) Pfitzer	T			+		
137	*Nervilia infundibulifolia* Blatt. & McCann	T	+		+		
138	*Nervilia plicata* (Andrews) Schltr.	T	+		+		+
139	*Oberonia brunoniana* Wight	E	+	+			+

SN	Name of taxon	Habit	AP	KA	OD	TE	TN
140	*Oberonia ensiformis* (Sm.) Lindl.	E	+	+	+		+
141	*Oberonia falconeri* Hook.f.	E	+		+		
142	*Oberonia gammiei* King & Pantl.	E			+		
143	*Oberonia mucronata* (D. Don) Ormerod & Seidenf.	E	+	+	+		
144	*Oberonia proudlockii* King & Pantl.	E		+	+		+
145	*Oberonia pyrulifera* Lindl.	E			+		
146	*Oberonia santapaui* Kapadia	E					+
147	*Oberonia verticillata* Wight	E					+
148	*Oberonia wightiana* Lindl.	E	+				+
149	*Odisha cleistantha* S. Misra	T	+		+		
150	*Pachystoma pubescens* Blume	T			+		
151	*Papilionanthe subulata* (Willd.) Garay	E		+			+
152	*Papilionanthe teres* (Roxb.) Schltr.	E	+		+		
153	*Pecteilis gigantea* (Sm.) Raf.	T	+		+		
154	*Pecteilis henryi* Schltr.	T			+		
155	*Pelatantheria insectifera* (Rchb.f.) Ridl.	E			+		
156	*Peristylus constrictus* (Lindl.) Lindl.	T			+		
157	*Peristylus goodyeroides* (D. Don) Lindl.	T	+		+		+
158	*Peristylus lawii* Wight	T	+		+		
159	*Peristylus parishii* Rchb.f.	T	+		+		
160	*Peristylus plantagineus* (Lindl.) Lindl.	T	+		+		
161	*Phalaenopsis deliciosa* Rchb.f.	E	+		+		
162	*Pholidota imbricata* Lindl.	E	+		+		
163	*Pholidota pallida* Lindl.	E	+				
164	*Phreatia elegans* Lindl.	E					+
165	*Pinalia polystachya* (A. Rich.) Kuntze	E					+
166	*Polystachya concreta* (Jacq.) Garay & H.R. Sweet	E	+	+	+		+
167	*Pomatocalpa decipiens* (Lindl.) J.J. Sm.	E			+		
168	*Pomatocalpa spicatum* Breda	E	+				

SN	Name of taxon	Habit	AP	KA	OD	TE	TN
169	*Porpax reticulata* Lindl.	E					+
170	*Rhynchostylis retusa* (L.) Blume	E	+	+	+		+
171	*Satyrium nepalense* D. Don	T	+				+
172	*Schoenorchis jerdoniana* (Wight) Garay	E					+
173	*Seidenfadeniella filiformis* (Rchb.f.) Christenson & Ormerod	E					+
174	*Sirhookera latifolia* (Wight) Kuntze	E	+				
175	*Smitinandia micrantha* (Lindl.) Holttum	E			+		
176	*Spiranthes sinensis* (Pers.) Ames	T	+		+		+
177	*Staurochilus ramosus* (Lindl.) Seidenf.	E	+		+		
178	*Taeniophyllum alwisii* Lindl.	E					+
179	*Tainia penangiana* Hook.f.	T			+		
180	*Taprobanea spathulata* (L.) Christenson	E	+				+
181	*Thunia alba* var. *bracteata* (Roxb.) N. Pearce & P.J. Cribb	E	+		+		
182	*Tropidia angulosa* (Lindl.) Blume	T	+		+		
183	*Tropidia hegderaoi* S. Misra	T					+
184	*Tropidia pedunculata* Blume	T			+		
185	*Vanda tessellata* (Roxb.) Hook. *ex* G. Don	E	+	+	+	+	+
186	*Vanda testacea* (Lindl.) Rchb.f.	E	+	+	+	+	+
187	*Vanilla walkeriae* Wight	V	+				
188	*Vanilla wightii* Lindl. *ex* Wight	V					+
189	*Zeuxine affinis* (Lindl.) Benth. *ex* Hook.f.	T			+		
190	*Zeuxine chowdheryi* Av. Bhattacharjee & Sabap.	T					+
191	*Zeuxine gracilis* (Breda) Blume	T			+		
192	*Zeuxine grandis* Seidenf.	T					+
193	*Zeuxine lindleyana* A.N. Rao	T			+		
194	*Zeuxine longilabris* (Lindl.) Trimen	T		+			+
195	*Zeuxine mooneyi* S. Misra	T			+		
196	*Zeuxine nervosa* (Wall. *ex* Lindl.) Benth. *ex* Trimen	T			+		
197	*Zeuxine strateumatica* (L.) Schltr.	T			+		
		Total	**89**	**27**	**128**	**17**	**84**

Endemic Orchids

Endemism in the flora of a geographical region provides an important insight into the biogeography. Endemism is therefore, rich in the orchids of Eastern Ghats. The Eastern Ghats region endemic orchids are distributed at different levels like, strict endemics, South India, Peninsular India and India.

The Peninsular Indian region is a part of Gondwana land plate and most of the endemics of this region are palaeoendemics. Highest degree of endemism is found in Western Ghats whereas lesser level in Eastern Ghats (Nayar, 1996). Jalal and Jayanthi (2012) reported about 15 species of *Bulbophyllum* and 11 species of *Dendrobium* are endemic to Peninsular India along with *Habenaria* (25 spp.) and *Oberonia* (17 spp.). The Eastern Ghats have some "ecological islands" that harbor 4% of endemic orchids (Jalal and Jeyanthi, 2012).

Strictly endemic to Eastern Ghats are 6 species namely, *Bulbophyllum panigrahianum* S. Misra, *Eria meghasaniensis* (S. Misra) S. Misra, *Habenaria panigrahiana* S. Misra, *Liparis espeevijii* S. Misra, *Zeuxine mooneyi* S. Misra, *Tropidia hegderaoi* S. Misra

Endemic to South India level

Bulbophyllum fischeri Seidenf., *B. kaitiense* Rchb.f., *B. tremulum* Wight, *Conchidium nanum* (A.Rich.) Brieger, *Dendrobium anamalayanum* Chandrab., V. Chandras. & N.C. Nair, *D. wightii* A.D. Hawkes & A.H. Heller, *Eria pauciflora* Wight, *Habenaria longicornu* Lindl., *Liparis biloba* Wight, *Luisia abrahamii* Vatsala, *L. birchea* Blume, *Malaxis intermedia* (A. Rich.) Seidenf., *Oberonia santapaui* Kapadia, *Pinalia polystachya* (A. Rich.) Kuntze, *Schoenorchis jerdoniana* (Wight) Garay, *Vanilla wightii* Lindl. *ex* Wight and and *Zeuxine chowdheryi* Av. Bhattacharjee & Sabap.

Endemic to Peninsular India level

Aerides maculosa Lindl., *Conchidium reticosum* (Wight) Ormerod, *Dendrobium aqueum* Lindl., *D. microbulbon* A. Rich., *D. nanum* Hook.f., *D. nodosum* Dalzell, *Habenaria crassifolia* A. Rich., *H. grandifloriformis* Blatt. & McCann, *H. heyneana* Lindl., *H. hollandiana* Santapau, *H. multicaudata* Sedgw., *H. ovalifolia* Wight, *H. rariflora* A. Rich., *Oberonia brunoniana* Wight and *O. proudlockii* King & Pantl.

Endemic to India

Dendrobium regium Prain, *Eulophia ochreata* Lindl., *Geodorum laxiflorum* Griff., *Odisha cleistantha* S. Misra and *Zeuxine lindleyana* A.N. Rao.

ORCRIDACEAE Juss.

Gen. Pl. 64. 1789 (4 Aug 1789) (*nom. cons*).

Type: *Orchis* L.

Perennial, but sometimes short-lived. Plants are terrestrial, epiphytic, lithophytic; autotrophic or rarely mycotrophic (saprophytic); habit: herbs, shrub-like or scrambling vines with rhizomes, tubers. Roots clustered or scattered along the stem or rhizome, with or without velamen, association with mycorrhizal fungi. Stems either sympodial or monopodial, usually leafy or rarely leafless, some genera base often swollen to form a pseudobulb; epiphytic species with aerial, photosynthesizing adventitious roots, often bearing 1 or more layers of dead cells (velamen). Leaves cauline or basal, 1 to many, alternate or opposite, distichous, plicate, convolute, duplicate or terete, frequently fleshy or leathery, sessile or petiolate, articulated or not, margin entire. Inflorescence lateral or terminal, axillary or not, erect to pendulous; racemose, spicate, subumbellate, capitate, paniculate, 1-to many-flowered. Flowers various sizes, often quite showy, usually zygomorphic, rarely actinomorphic, bisexual, sessile or pedicellate, resupinate or not. Sepals usually free, but sometimes variously adnate; dorsal sepal often dissimilar to lateral sepals, lateral sepals sometimes adnate to a column foot to form a mentum. Petals free or rarely partly adnate to sepals, similar to sepals or not; median tepal of the inner whorl often transformed into labellum/lip. Labellum entire, variously lobed or 2-or 3-partite, ornamented or not, with calli, ridges, hairs or crests, with or without a basal spur or nectary. Column part formed by the fusion of style and stigma, short to long, with or without a basal foot, occasionally winged or with lobes or arms at apex or ventrally. Stigma single or 2 or 3-lobed, concave or convex, wet, entire or spilt into three parts. The midlobe of stigma often modified to form a rostellum, a structure of various shapes and sizes. Anther mostly 1, less often 2 or 3, terminal or ventral on column, caplike or opening by longitudinal slits, erect, reflexed or incumbent, fixed or movable or detachable. Pollen usually forming distinct pollinia; pollinia 2, 4, 6 or 8, mealy, waxy or horny, compact or sectile, sessile or attached by stalks (caudicles or stipes) to 1 or 2 sticky viscidia. Ovary inferior, 1-locular, placentation parietal (or rarely 3-locular and placentation axile). Fruit a capsule, rarely berrylike. Seeds numerous, dustlike, lacking endosperm, rarely winged, adapted to anemochory.

SYSTEMATIC ENUMERATION

Key to Genera

1a. Plants terrestrial 2
1b. Plants epiphytic or lithophytic 30
2a. Plants saprophytic without any green parts 3
2b. Plants with green parts (autotrophic) 5
3a. Sepals and petals fused and form 5-lobed tube .... **Didymoplexis**
3b. Sepals and petals free 4
4a. Roots fasciculate; lip clawed, not spurred .... **Aphyllorchis**
4b. Root tuberous; lip not clawed, spurred .... **Epipogium**
5a. Root climbers; stem sub-cylindrical, branching; leaves reduced to scales .... **Vanilla**
5b. Plants not as above 6
6a. Plants with root-stem tuberoids or corms or pseudobulbous 13
6b. Plants not as above 7
7a. Flowers arranged in a spirally twisted spike .... **Spiranthes**
7b. Flowers not spirally arranged 8
8a. Plants with decumbent rhizome 9
8b. Plants erect, without decumbent rhizome 12
9a. Sepals connate into a tube for half of their length .... **Cheirostylis**
9b. Sepals free 10
10a. Mesochile (claw) of the lip with 5-8 pairs long fimbriate on either side .... **Anoectochilus**
10b. Lip not fimbriate 11
11a. Sac of the lip with glands; stigma two-lobed .... **Zeuxine**
11b. Sac of the lip hairy inside; stigma entire .... **Goodyera**
12a. Plants about 1.5 m high; inflorescence branched; flowers more than 2 cm long .... **Corymborkis**
12b. Plants less than 50 cm high; inflorescence simple; flowers less than 1 cm long .... **Tropidia**
13a. Inflorescence hook-shaped (decurved) .... **Geodorum**

13b. Inflorescence not hook-shaped and erect... 14
14a. Sepals connate and form gibbous tube... ....**Acanthephippium**
14b. Sepals free....................... 15
15a. Pollinia 8 16
15b. Pollinia 4 or 2................. 18
16a. Mid-lobe of lip bilobuled..................... ....**Calanthe**
16b. Midlobe of lip not bilobuled....... 17
17a. Leaf single; inflorescence with leaves; lip with spur...................................... **Tainia**
17b. Leaves two or more; inflorescence appears without leaves; lip with saccate base.......... ....**Pachystoma**
18a. Lip with two spurs........ ....**Satyrium**
18b. Lip without or with single spur...19
19a. Flowers non-resupinate... 20
19b. Flowers resupinate..22
20a. Lip with two large basal auricles............. ....**Crepidium**
20b. Lip without any basal auricles....... 20
21a. Lip not lobed, apex pectinate...**Malaxis**
21b. Lip 3-lobed, apex entire... .**Dienia**
22a. Plants with single leaf... 23
22b. Plants with two or more leaves.. 24
23a. Plants hysteranthous; leaves broad............. ....**Nervilia**
23b. Plants not as above; leaves lanceolate... .**Chrysoglossum**
24a. Pollinia 4.................... ....**Liparis** ***p.p.***
24b. Pollinia 2...25
25a. Root stem tuberoid; inflorescence terminal............. 26
25b. Plants with corms or pseudobulbs; inflorescence lateral.............................**Eulophia**
26a. Lip without spur and bilobed..**Disperis**
26b. Lip with perfect spur, entire or 3-lobed............... 27
27a. Flowers large, 8-10 cm across; lateral lobes of lip pectinate..........................**Pecteilis**
27b. Flowers small, less than 8 cm across; lateral lobes of lip not pectinate 28
28a. Ovary close to the rachis, erect; lip with short spur; stigma sessile..............**Peristylus**
28b. Ovary spreading; lip with long spur; stigma stalked 29
29a. Flowers cleistanthus; spur saccate, laterally compressed .... **Odisha**
29b. Flowers not as above; spur elongate, tubular .... **Habenaria**
30a. Plants without leaves and stem, with photosynthetic roots..................................... 31
30b. Plants with leaves and stem, without photosynthetic roots 32
31a. Sepals and petals freely spreading; lip without spur................................**Chiloschista**
31b. Sepals and petals united into tube; lip with spur ....**Taeniophyllum**

32a. Plants acaulescent; leaves equitant............ ....**Oberonia**

32b. Plants caulescent; leaves not equitant.. 33

33a. Lip with two prominent spurs............. ....**Diplocentrum**

33b. Lip with single spur or not 34

34a. Lip appears like a bee...**Cottonia**

34b. Lip not as above 35

35a. Leaves terete (cylindrical) 36

35b. Leaves dorsiventral 39

36a. Lip lacking spur; inflorescence much shorter than the leaves ....**Luisia**

36b. Lip with spur; inflorescence little shorter or longer than the leaves......................... 37

37a. Sepals and petals yellow-green with maroon veins; pollinia 4..........**Cleisostoma *p.p.***

37b. Sepals and petals not as above; pollinia 2 38

38b. Flowers orange-yellow..**Seidenfadeniella**

38b. Flowers not as above ....**Papilionanthe**

39a. Plants with sympodial growth 40

39b. Plants with monopodial growth... 54

40a. Pollinia 8 41

40a. Pollinia 4 or 2 47

41a. Plants without pseudobulbs ....**Thunia**

41b. Plants with pseudobulbs 42

42a. Sepals united and form sub-campanulate tube ....**Porpax**

42b. Sepals not united and free 43

43a. Pseudobulbs compact, depressed, discoid or orbicular or globose.......... **Conchidium**

43b. Pseudobulbs not as above.. 44

44a. Flowers pale brown with darker red veins and markings ....**Callostylis**

44b. Flowers not as above 45

45a. Floral bracts deflexed .... **Pinalia**

45b. Floral bracts not deflexed.. 46

46a. Pollinia joined by a long caudicle...**Phreatia**

46b. Pollinia not joined by a caudicle ....**Eria**

47a. Flowers bifarious..**Pholidota**

47b. Flowers not bifarious.. 48

48a. Pollinia 2 ....**Cymbidium**

48b. Pollinia 4 49

49a. Panicle inflorescence.............. 50

49b. Simple inflorescence 51

50a. Flowers non-resupinate................... ....**Polystachya**

50b. Flowers resupinate ....**Sirhookera**

Distribution: ***World*:** Bangladesh, Bhutan, India, Laos, Myanmar, Nepal, Thailand, and Vietnam. ***India:*** Andhra Pradesh, Arunachal Pradesh, Assam, Bihar, Manipur, Meghalaya, Mizoram, Odisha, Sikkim, Tripura, Uttar Pradesh, Uttarakhand and West Bengal. ***Eastern Ghats:*** *AP:* Minumuluru forest; *OD:* Gopalpur-Gunupur forest, Govindapali-Gupteswara forest and Similipal forest.

Specimens Examined: *AP:* Minumuluru, Wongasara coffee plantation (VSKP), *SM* 98 (RPRC). *OD:* Similipal forest (MB), *SM* 905 (CAL).

Acampe ochracea (Lindl.) Hochr., Bull. New York Bot. Gard. 6: 270. 1910; Ramakrishna in Pullaiah, Fl. Andhra Pr. 3: 931. 1997; Misra *et al.,* Bull. Bot. Surv. India 50: 133. 2008; Misra, Orchids Orissa 577. 2004. *Saccolabium ochraceum* Lindl., Edwards's Bot. Reg. 28 (Misc.): 2. 1842; Hook. f., Fl. Brit. India 6: 62. 1890.

Type: Sri Lanka: *Horton s.n.* (K-LINDL).

Etymology: The specific epithet (Latin) reference to the flower colour of the species.

Epiphytic herbs, *ca* 30 cm high. Stem scandent, tufted, terete. Leaves coriaceous, sessile, linear-oblong, unequally bilobed, *ca* 20 × 2.5 cm. Inflorescence leaf-opposed, as long as or longer than the leaves, *ca* 25 cm long; panicels laxly many-flowered. Floral bracts broadly triangular, acute, persistent, *ca* 1.6 × 1.8 mm. Pedicel with ovary pale green, *ca* 6 mm long. Flowers fragrant, *ca* 1 cm across. Sepals and petals light yellow to ochre with irregular maroon transverse markings, apex obtuse; dorsal sepal broadly oblanceolate, *ca* 5.5 × 2 mm; lateral sepals oblong-oblanceolate, *ca* 5 × 2.2 mm; petals spathulate, *ca* 4.5 × 1.5 mm. Lip white, 3-lobed; laterals erect, broadly triangular, *ca* 1 mm long; midlobe large, deflexed, oblong-ovate or broadly oblong, *ca* 3 × 1.5 mm, obtuse or rounded, surface warty, margins crispate or undulate; spur cylindric, *ca* 4 mm long, hairy inside. Column yellowish-maroon, truncate, *ca* 2 × 1.2 mm, with two terminal horns. Stigma suborbicular; rostellum trifid. Pollinia 4, globose-ovoid, 1 × 0.8 mm. Capsules oblong-fusiform, *ca* 3.5 cm long.

Fl. & Fr.: December-February.

Habitat: Rare in moist deciduous forest (500-800m).

Distribution: ***World*****:** Bhutan, Cambodia, China, India, Laos, Myanmar, Nepal, Sri Lanka, Thailand and Vietnam. ***India:*** Andhra Pradesh, Arunachal Pradesh, Karnataka, Kerala, Meghalaya, Mizoram, Nagaland, Odisha, Sikkim, Tamil Nadu and W. Bengal. ***Eastern Ghats:*** *AP*: Minumuluru forest and Ebul Reserve forest; *OD:* Similipal forest.

Specimens Examined: *AP:* Maredumilli (EG), *KS* 23464 (SKU); Minumuluru (VSKP), *GVS* 29601 (MH); Minumuluru (VSKP), *SM* 99 (RPRC); Ebul Reserve forest (EG), *VNS* 570 (CAL). *OD:* Similipal forest-Barigam (MB), *SM* 1042 (CAL); Similipal forest-Chahala (MB), *SM* 67 (CAL); Simipal-Barhakamuda (MB), *SM* 82 (CAL).

Acampe praemorsa (Roxb.) Blatt. & McCann, J. Bombay Nat. Hist. Soc. 35: 495. 1932; Ramakrishna in Pullaiah, Fl. Andhra Pr. 3: 931. 1997; Misra *et al.,* Bull. Bot. Surv. India 50: 133. 2008; Misra, Orchids Orissa 584. 2004; Seidenfaden in K.M. Matthew, Fl. Tamilnadu Carnatic 3: 1602. 1983. *Epidendrum praemorsum* Roxb. in Pl. Corom. 1: 34, t. 43. 1795. *Acampe wightiana* (Lindl. *ex* Wight) Lindl., Fol. Orchid. 4: 2. 1853; C.E.C. Fisch., Fl. Madras 3: 1447. 1928. *Saccolabium praemorsum* (Roxb.) Hook.f., Fl. Brit. India 6: 62. 1890 (*nom. illeg*); Haines, Bot. Bihar & Orissa, 1180. 1924; Mooncy, Suppl. Bot. Bihar & Orissa 215. 1956.

Type: India: Andhra Pradesh, Circar mountains, Roxburgh *icon*. 43. 1795.

Etymology: The specific epithet refers to the papillose nature of the lip.

Epiphytic or lithophytic herbs, *ca* 45 cm high. Stem erect or pendulous, stout, terete. Leaves coriaceous, sessile, lorate or linear-oblong, unequally bilobed, *ca* 25 × 3.5 cm. Inflorescence leaf-opposed or supra-axillary, shorter than the leaves, simple or rarely branched, *ca* 10 cm long; corymbs densely few-flowered. Floral bracts green with maroon dots, broadly triangular, persistent, *ca* 2.5 × 2.5 mm. Pedicel with ovary green with maroon dots, *ca* 1 cm long. Flowers fragrant, fleshy. Sepal and petals pale yellow with irregular transverse maroon bands, apex obtuse; dorsal sepal broadly oblong-oblanceolate, *ca* 1.2 × 0.5 cm; lateral sepals falcate, oblong-oblanceolate, *ca* 1 × 0.6 cm; petals falcate, spathulate, *ca* 1 × 0.3 cm. Lip white with maroon line at the base, deflexed, 3-lobed; base hairy within, with a small rounded sac; laterals narrowly rounded; midlobe large, truncately-ovate, margin irregularly notched, surface papillose, *ca*

5 × 3.5 mm. Column yellow with maroon blotches, stout, oblong-suborbicular, *ca* 3 × 3 mm, with two minute horns. Stigma suborbicular; rostellum bifid. Pollinia 2, oblong-globose, 1 × 0.8 mm. Capsules oblong or fusiform, *ca* 6 cm long.

Fl. & Fr.: May-October.

Habitat: Common on tree branches and rocks in moist places of moist deciduous open forests (above 400 m).

Distribution: ***World*:** Bhutan, India, Myanmar, Seychelles and Sri Lanka. ***India:*** Andhra Pradesh, Arunachal Pradesh, Bihar, Gujarat, Karnataka, Kerala, Madhya Pradesh, Maharashtra, Odisha, Sikkim, Tamil Nadu and West Bengal. ***Eastern Ghats:*** Occasionally found in Eastern Ghats of Andhra Pradesh, Karnataka, Odisha and Tamil Nadu.

Specimens Examined: *AP:* Nagaluty (KNL), *BR* & *KP* 34619 (SKU); Matta (KNL), *BSS* & *KP* 32347 (SKU); Rollapenta (PKM), *BR* & *KP* 34622 (SKU); Chittoor, *PS* 2412 (PCM); Near Maredumilli (EG), *KS* 23464 and *BR* & *KNR* 10618 (SKU); Bhojeripakala, Maredumilli (EG), *MM* 101325 (BSID); Ettukkuru, Maredumilli (EG), *MM* 105139 (BSID); Muzumamidivalasa–Maredumilli (EG), *MM* 101221 (BSID); Way from Maredumilli to Rampachodavaram (EG), *GVS* 27407 (MH); Lakshmipuram (VSKP), *CAB* 1794 (MH); Chintapalli towards Narsipatnam (VSKP), *GVS* 30077 (MH); Wongasava (VSKP), *SM* 0100 (RPRC); Jarda (VZN), *MV* 19837 (AU); Near Gummada (SKLM), *GVS* 62439 (CAL); Papikonda R.F., *D.C.S. Raju* 503 (CAL). *OD:* Pandakhola (GJM), *J.K. Das* & *U.P. Samader* 26314 (CAL); Jhumal Rampuri (KH), *B. Sabui* & *Party* 13551 (CAL); Konda Kamberu (KP), *D.C.S. Raju* 847

(CAL). *TN*: Kolli hills (NK), *KMM & CR* 2107, 4814 (RHT); Arapaleeshwarar falls, Kolli hills (NK), *AD* 22869 (RHT), *GP* 24186 (RHT).

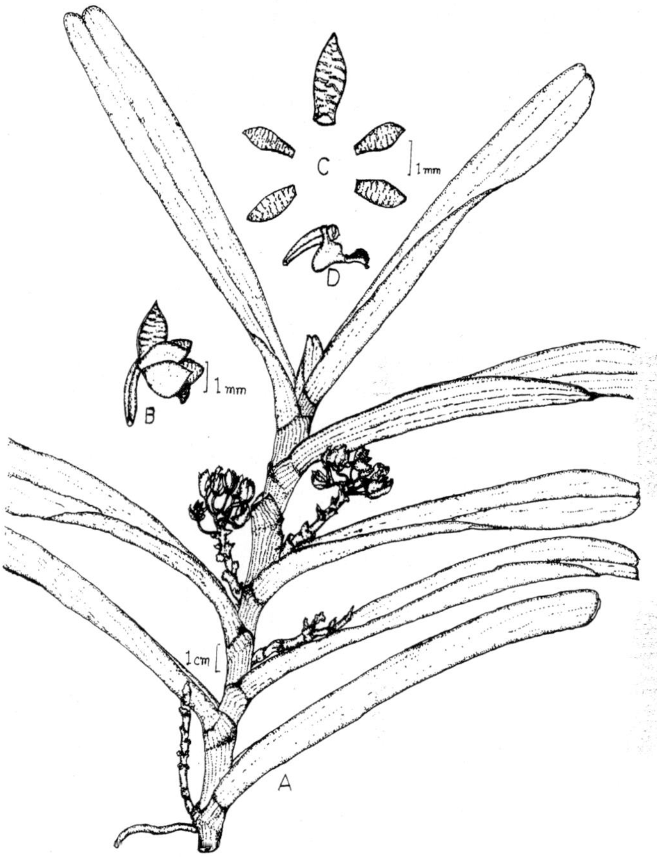

Figure 1. **Acampe praemorsa** (Roxb.) Blatter & McCann
A. Twig; B. Flower; C. Sepals & Petals; D. Lip

Acampe rigida (Buch.-Ham. *ex* Sm.) P.F. Hunt, Kew Bull. 24: 98. 1970; Misra, Orchids Orissa 587. 2004; Misra *et al.,* Bull. Bot. Surv. India 50: 133. 2008. *Aerides rigida* Buch.-Ham. *ex* Sm., Cycl. 39: 12. 1818. *Saccolabium longifolium* (Lindl.) Hook.f., Fl. Brit. India 6: 62. 1890.

Type: Nepal: *sine. loc.*, *Buchanan-Hamilton s.n.* (holo: LINN).

Etymology: The specific epithet refers to the nature of inflorescence of the species.

Epiphyte or lithophyte, robust herbs, *ca* 1 m high. Stem stout, terete, branched or unbranched. Leaves coriaceous, sessile, oblong or lorate, unequally bilobed, *ca* 40 × 4 cm. Inflorescence shorter than the leaves, simple or rarely branched, *ca* 20 cm long; racemes laxly-flowered. Floral bracts broadly triangular, *ca* 3 × 3.2 mm, persistent. Pedicel with ovary yellowish-green, *ca* 1.5 cm long. Flowers fragrant. Sepals and petals golden yellow banded with maroon blotches, apex rounded or obtuse; dorsal sepal obovate-elliptic, *ca* 1.2 × 0.6 cm; lateral sepals obovate, *ca* 1.2 × 0.6 cm. Petals falcate, oblong-oblanceolate, *ca* 1.2 × 0.5 cm. Lip white with purple bands, 3-lobed; laterals rectangular, *ca* 4 mm long, rounded; midlobe deflexed, broadly ovate, entire, surface covered with hairs, *ca* 6 × 5 mm; spur conical, *ca* 3 mm long, densely hairy inside. Column ovate-triangular, *ca* 4 × 5 mm, with two horns. Stigma sub-orbicular. Pollinia 2, globose-ovoid, 1 × 0.7 mm. Capsules linear-oblong, *ca* 6 cm long.

Fl. & Fr.: August-October.

Habitat: Rare in semi-evergreen and moist deciduous forests (above 600 m).

Distribution: ***World*:** Africa, Bhutan, Cambodia, China, Comoro Island, India, Laos, Madagascar, Malaysia, Myanmar, Nepal, Philippines, Sri Lanka, Taiwan, Thailand and Vietnam. ***India:*** Andhra Pradesh, Arunachal Pradesh, Assam, Goa, Karnataka, Kerala, Maharashtra, Manipur, Meghalaya, Nagaland, Odisha, Sikkim, Tamil Nadu, Uttarakhand and West Bengal. ***Eastern Ghats:*** *AP:* Maredumilli forest; *OD:* Similipal forest.

Specimens Examined: *AP:* Bodhuluru (EG), *SM* 125 (RPRC); Maredumilli (EG), *V.S. Raju* 6000 (KUH). *OD:* Similipal forest-Champaghat (MB), *SM* 1305 (CAL); Similipal forest-Jenabila (MB), *SM* 848 (CAL).

Acanthephippium Blume *ex* Endl.

Gen. Pl. (Endlicher) 200. 1837.

World:13/India:3/EG:2

Type: *Acanthephippium javanicum* Blume

Etymology: The generic name is derived from the Greek words, *acantha* (thorn) and *ephippion* (saddle), refers to the shape of the lip (saddle).

Distribution: Indian subcontinent to China, Japan, Southeast Asia, New Guinea and the southwest Pacific.

Key to the species

1a. Leaves broadly elliptic or ovate-oblong; disc with 3 keel-shaped lamellae, with papillary ridges..**A. bicolor**

1b. Leaves oblong-elliptic or elliptic-lanceolate; disc with 3-5 keel-shaped lamellae, coarse ridges..**A. sylhetense**

Acanthephippium bicolor Lindl., Edwards's Bot. Reg. 20: t. 1730. 1835; Hook.f., Fl. Brit. India 5: 815. 1888; C.E.C. Fisch., Fl. Madras 1426. 1928;

Misra, Orchids Orissa 487. 2004; Seidenfaden in K.M. Matthew, Fl. Tamilnadu Carnatic 3: 1572. 1983.

Lectotype: Sri Lanka: Kandy district, *Thwaites* 2365 (PDA; designated by Jayaweera, 1981).

Etymology: The specific epithet refers to the bicolour nature of the perianth.

Terrestrial robust herbs, *ca* 50 cm high. Pseudobulbs crowded, green, smooth, oblong-ovate, nodded, with persisting sheaths, *ca* 15 × 5 cm. Leaves 2 or 3, plicate, broadly elliptic or ovate-oblong, acuminate, *ca* 50 × 11 cm; petiole *ca* 2 cm long, sheathing. Inflorescence lateral, much shorter than the leaves, sheathing with large bracts, *ca* 10 cm long; racemes laxly 3-7-flowered. Floral bracts as long as or longer than the ovary, pale, ovate-lanceolate, acute, *ca* 3 × 2 cm. Pedicel and ovary whitish-rose, *ca* 4 cm long. Flowers like gibbous-based jugs, yellowish-white with pink dots, fleshy. Sepals connate for about half the length and forming a gibbous tube, mouth oblique, tips recurved; lobes oblong, obtuse, 7-veined, *ca* 3 × 2 cm. Petals included in sepaline-tube, erect, clawed, oblong-lanceolate or ovate, obtuse, 5-veined, *ca* 3 × 1.4 cm. Lip included sepaline-tube, stipitate, mobile, *ca* 1.6 × 1.2 cm, 3-lobed; laterals hatchet-shaped; midlobe strongly reflexed, tongue-shaped, ovate, obtuse, grooved; disc with 3 coarsely dentate keels, with pair of slender papillose ridges. Column whitish-yellow, sub-cylindric, *ca* 1.5 × 0.7 cm, winged; foot *ca* 4 mm long. Stigma broadly orbicular. Pollinia 8, oblong-ovoid or ovoid, 4 × 2 mm. Capsules ellipsoid, *ca* 8 cm long.

Fl. & Fr.: May-July.

Habitat: Rare along the perennial streams under shady conditions in humus-rich soils of semi-evergreen and moist deciduous forests (above 800 m).

Distribution: ***World:*** India and Sri Lanka. ***India:*** Karnataka, Kerala, Odisha and Tamil Nadu. ***Eastern Ghats:*** *AP:* Galikonda hills; *OD:* Similipal forest; *TN*: Kolli hills and Shevaroy hills.

Specimens Examined: *AP*: Galikonda hills (VSKP), *KP* 6422 (BSID). *TN*: Kolli hills, Ariyur shola (NK), *SKS* 1105 (SGH); Kolli hills, Solakkaud (NK), *N. Ahmed Sherif* & party *s.n.* (MH); Kolli hills (NK), *KMM* 4633, 25350, 25392, 63607 (RHT); Kolli hills (NK) *KMM* & *SJB* 30191 (RHT); Kolli hills, Mathiketton shola (NK) *KMM* 28279, 30442 (RHT); Kolli hills, Madal Shola (NK), *KMM, SJB* & *NRR* 28900 (RHT); Kolli hills (NK), *PP* & *CM* 21493 (RHT); Shevaroys hills, Yercaud-Shervaroyan Temple (SLM), *DIA* 7838 (RHT); Shevaroys hills, Kakashola (SLM), *NV* & *TSJ* 22443 (RHT); Shevaroys hills, Kakashola (SLM), *KMM* 23368 (RHT); Shevaroys hills, Yercaud (SLM), *KMM* 62006 (RHT); Shevaroys hills, bauxite mines shola (SLM), *AVNR* 23069 (MH); Shevaroy hills (SLM), *E. Barnes* s.n. (DD); Sothuparai (NK), *KMM* 30428, 62247, 63752 (RHT); Peria kada aar (NK), *KMM* 63068 (RHT).

Acanthephippium sylhetense Lindl., Gen. Sp. Orchid. Pl. 177. 1831; Hook.f., Fl. Brit. India 5: 815. 1888; Misra, Orchids Orissa 490. 2004.

Type: Bangladesh: Sylhet, *Wallich s.n.* (not located).

Etymology: This species was named after the type locality, Sylhet (Bangladesh).

Terrestrial herbs. Pseudobulbs narrow-conical or ovoid-cylindrical. Leaves 2 or 3, petioled, oblong-elliptic or elliptic-lanceolate, *ca* 40 cm long. Inflorescence lateral; racemes laxly 3-6-flowered. Floral bracts ovate-lanceolate. Pedicel with ovary *ca* 4 cm long. Flowers like gibbous-based jugs, yellowish and dotted with maroon or red. Sepals connate and forming gibbous tube, mouth oblique, tips recurved; lobes oblong, *ca* 4.5 cm long. Petals included in sepaline-tube, ovate. Lip included in sepaline-tube, 3-lobed; laterals hatchet-shaped; midlobe tongue-shaped; disc with 3-5-toothed keel-shaped lamellae. Capsules oblong-ellipsoid, *ca* 6 cm long.

Fl. & Fr.: May-October.

Habitat: Rare along perennial streams under dense shady conditions in humus rich soils of evergreen and semi-evergreen forests (above 800 m).

Distribution: ***World***: Bangladesh, Bhutan, China, Japan, Malaysia, Myanmar, Philippines, Ryukyu Is., Taiwan and Thailand. ***India:*** Arunachal Pradesh, Assam, Meghalaya, Odisha and Sikkim. ***Eastern Ghats:*** *OD:* Similipal forest.

Specimens Examined: *OD*: Similipal forest-Meghasini (MB), *SM* 101 (CAL); Similipal forest-Garandia nala (MB), *SM* 1266 & 916 (CAL).

Aerides Lour.
Fl. Cochinch. 525. 1790.
World:27/India:9/EG:4

Type: *Aerides odorata* Lour.

Etymology: The generic name is derived from the Greek words, *aer* (air) and *idus* (nature), a reference to the epiphytic habit.

Distribution: Tropical and Subtropical Asia.

Key to the species

1a. Lateral lobes of lip broader than the midlobe; spur hook-shaped................. **A. odorata**

1b. Lateral lobes of the lip smaller than the midlobe; spur not as above............................ 2

2a. Midlobe of lip oblong ovate; racemes perfectly erect ....**A. ringens**

2b. Midlobe of lip hastate-ovate or broadly ovate; racemes pendulous 3

3a. Inflorescence simple; spur *ca* 4 mm long, not hooked .... **A. multiflora**

3b. Inflorescence simple or branched; spur more than 5 mm long, hooked.....**A. maculosa**

Aerides maculosa Lindl., Edwards's Bot. Reg. 31: 58. 1845; Hook.f., Fl. Brit. India 6: 45. 1890; C.E.C. Fisch., Fl. Madras 1442. 1928; Haines, Bot. Bihar & Orissa, 1182. 1924; Pullaiah & Karuppusamy, Fl. Andhra Pr. 5: 67. 2008; Misra, Orchids Orissa 660. 2004.

Iconotype: India: Tamil Nadu, Courtellam, Rollisons (icon).

Etymology: The specific epithet refers to the purple spots nature of the perianth.

Epiphytic herbs, *ca* 8 cm high. Stem erect or pendulous. Leaves coriaceous, sessile, linear-oblong, unequally lobed, specked with purple spots, *ca* 25 × 3 cm. Inflorescence axillary, simple or branched, *ca* 30 cm long, densely many-flowered racemes. Floral bracts greenish-purple, triangular, acute, *ca* 2 × 2 mm. Pedicel with ovary whitish-rose, *ca* 1 cm long. Flowers fragrant, whitish-rose and sepals and petals with purple spots. Dorsal sepal oblong-obovate, rounded, 7-8-veined, *ca* 1 × 0.5 cm; lateral sepals broadly obovate, rounded, *ca* 1 × 0.8 cm. Petals falcate, obovate, rounded, 3-veined, *ca* 1 × 0.4 cm. Lip spreading, clawed, *ca* 2 cm long, 3-lobed; laterals quadrate, rounded, with callus; midlobe incurved, broadly ovate, feebly retuse, margins crenate, with longitudinal deep purple ridges, *ca* 1.2 × 1.2 cm; spur purplish-white, cylindrical, hooked, *ca* 8 mm long. Column purplish-white, stout, sub-cylindric, *ca* 6 × 3 mm; foot *ca* 4 mm long. Stigma sub-orbicular. Pollinia 2, sub-globose, *ca* 1.8 × 0.8 mm. Capsules obovoid, *ca* 2 cm long.

Fl. & Fr.: June-July.

Habitat: Rare in moist deciduous forests (600-800 m).

Distribution: Endemic to India (Andhra Pradesh, Chhattisgarh, Gujarat, Jharkhand, Karnataka, Kerala, Maharashtra, Odisha, Rajasthan and Tamil Nadu). ***Eastern Ghats:*** *AP:* Nallamala hills; *OD*: Govindapali forest.

Specimens Examined: *AP:* Rollapenta (PKM), *BR* & *KP* 34620 & 34621 (SKU); Nallamalais (KNL), *VBH* 83958, 84926 (MH, BSID). *OD:* Jaypur (KP), *SM* 13 (CAL).

Aerides multiflora Roxb., Pl. Coromandel 3: 68. 1820; Hook.f., Fl. Brit. India 6: 44. 1890; Haines, Bot. Bihar & Orissa, 1182. 1924; Ramakrishna in Pullaiah, Fl. Andhra Pr. 3: 932. 1997; Misra *et al.,* Bull. Bot. Surv. India 50: 133. 2008; Misra, Orchids Orissa 654. 2004.

Iconotype: India or Bangladesh? near Sylhet, *Roxburgh* icon 2346.

Etymology: The specific epithet refers to the many-flowered raceme.

Epiphytic herbs, *ca* 10 cm high. Stem erect or pendulous. Leaves coriaceous, keeled, sessile, linear-oblong, obliquely bifid, *ca* 34 × 2 cm. Inflorescence axillary, simple, pendulous, *ca* 40 cm long; racemes densely many-flowered. Floral bracts light purple, triangular-ovate, acute, *ca* 4 mm long. Pedicel with ovary whitish-rose, *ca* 5 mm long. Flowers fragrant, purplish-white. Dorsal sepal oblong, rounded or obtuse, 5-veined, *ca* 8 × 4 mm; lateral sepals broadly obovate or ovate, rounded or obtuse, *ca* 8 × 6 mm. Petals oblong-elliptic or oblong-obovate, obtuse, 5-veined, *ca* 8 × 3.5 mm. Lip clawed, inflexed, 3-lobed; laterals decurrent on the column, subquadrate or semi-circular, margins crenulated, *ca* 2 × 2 mm; midlobe hastate-ovate, margins irregularly crenulate and recurved, base with callus, *ca* 1 × 4 mm; spur broadly conical, compressed, *ca* 4 mm long. Column whitish-purple, sub-cylindric, *ca* 3 × 2 mm; foot *ca* 3 mm long. Stigma orbicular. Pollinia 2, sub-globose, 0.8 × 0.8 mm. Capsules fusiform, *ca* 4 cm long.

Fl. & Fr.: May-July.

Habitat: Occasionally on tree trunks and branches in semi-evergreen and moist deciduous forests (600-1000 m).

Distribution: ***World*****:** Bangladesh, Bhutan, Cambodia, India, Java, Laos, Malaysia, Myanmar, Nepal, Thailand and Vietnam. ***India:*** Andaman & Nicobar Is., Andhra Pradesh, Arunachal Pradesh, Assam, Chhattisgarh, Himachal Pradesh, Jharkhand, Madhya Pradesh, Manipur, Meghalaya, Mizoram, Nagaland, Odisha, Rajasthan, Sikkim, Tamil Nadu, Tripura, Uttarakhand and West Bengal. ***Eastern Ghats:*** Occasionally found in Eastern Ghats.

Specimens Examined: *AP:* Yarlagadda (EG), *SS* 626 (AU); way to Sesharayi to Nulakamaddi (EG), *VNS* 282 (CAL); Donkarayi to Mangapadu (EG), *GVS* 29688 (MH). *OD:* Jaypur (KP), *SM* 12 (CAL); Bagghola forest, *GVS s.n.* (CAL).

Aerides odorata Lour., Fl. Cochinch. 525. 1790; Hook.f., Fl. Brit. India 6: 47. 1890; C.E.C. Fisch., Fl. Madras 1443. 1928; Haines, Bot. Bihar & Orissa, 1182. 1924; Mooney, Suppl. Bot. Bihar & Orissa 215. 1956; Ramakrishna in Pullaiah, Fl. Andhra Pr. 3: 932. 1997; Misra *et al.*, Bull. Bot. Surv. India 50: 133. 2008; Misra, Orchids Orissa 649. 2004.

Types: Vietnam (Cochinchina): *Loureiro s.n.* (BM & K (images!)).

Etymology: The specific epithet refers to the scented nature of flowers.

Epiphytic herbs, *ca* 60 cm high. Stem erect or pendent. Leaves coriaceous, keeled, sessile, lorate or linear-oblong, unequally bilobed, *ca* 25 × 3.5 cm. Inflorescence extra-axillary, simple, deflexed, *ca* 30 cm long; racemes densely many-flowered. Floral bracts whitish-purple, linear-lanceolate or triangular, acute, 3-veined, persistent, *ca* 5 mm long. Pedicel with ovary white, *ca* 1.5 cm long. Flowers sweet-scented, purplish-white. Dorsal sepal oblong, acute, 7-9-veined, *ca* 1.2 × 0.6 cm; lateral sepals broadly triangular-ovate or oblong-ovate, acute, *ca* 1.4 × 0.8 cm. Petals falcate, oblong-lanceolate, acute, 6-veined, *ca* 1.1 × 0.6 cm. Lip almost enclosing the column, sessile, 3-lobed; laterals erect, triangular, margins crenulated, *ca* 6 × 5.5 mm; midlobe erect, lying between the edge of the lateral lobes, incurved, oblong-lanceolate, *ca* 6 × 2.8 mm; spur hooked, laterally compressed, infundibuliform, *ca* 1.5 cm long. Column whitish, oblong-ovate, *ca* 4 × 2 mm, beaked; foot *ca* 8 mm long. Stigma rectangular. Pollinia 2, globose-ovoid, 2 × 0.8 mm. Capsules oblong-clavate, *ca* 2 cm long.

Fl. & Fr.: April-June.

Habitat: Occasional in semi evergreen and moist deciduous forests (600-1200 m).

Distribution: *World*: Bangladesh, Bhutan, Borneo, China, India, Indonesia, Java, Laos, Malaysia, Moluccas, Myanmar, Nepal, Philippines, Sulawesi, Sumatra, Thailand, Timor and Vietnam. ***India:*** Andaman Is., Andhra Pradesh, Arunachal Pradesh, Assam, Bihar, Jharkhand, Karnataka, Manipur, Meghalaya, Mizoram, Nagaland, Odisha, Sikkim, Tamil Nadu, Tripura, Uttar Pradesh, Uttarakhand and West Bengal. ***Eastern Ghats:*** Common in Eastern Ghats.

Specimens Examined: *AP:* Maredumilli (EG), *GVS* 24293 (MH); Sesharayi Hills (EG), *VNS* 626 (CAL); Sesharayi to Nulakamaddi (EG) *VNS* 282 (CAL); Yarlagadda (EG), *VNS* 128 (CAL); Sesharayi Hills (EG), *SS* 626 (AU); Chinthiru (EG), *MM* 100716; Anantagiri-Galikonda (VSKP), *BR* & *KP* 34645, 34646 (SKU); Near Gorapur station (VSKP), *GVS* 44406 (MH); Tanjavanam (VSKP), *GVS* 42611 (MH); Lakshmipuram (VSKP), *CAB* 1795 (MH); G. Madugula (VSKP), *GVS* 47221 (MH). *OD:* Boragharo (GJM), *VNS* 5951 (MH); Rebana forest-Polaspal (KJ), *G. Panigrahi* 8468 (CAL); Ramogiri, *GVS* 30325 (CAL); Near Kohla village, *G. Panigrahi* 8240 (CAL).

Aerides ringens (Lindl.) C.E.C. Fisch., Kew Bull. 284. 1928 *et* Fl. Madras 1442. 1928; Ramakrishna in Pullaiah, Fl. Andhra Pr. 3: 933. 1997; Misra *et al.*, Bull. Bot. Surv. India 50: 133. 2008; Misra, Orchids Orissa 656. 2004. Seidenfaden in K.M. Matthew, Fl. Tamilnadu Carnatic 3: 1593. 1983. *Saccolabium ringens* Lindl., Gen. Sp. Orchid. Pl. 221. 1833. *Aerides radicosa* A. Rich., Ann. Sci. Nat., Bot. II, 15: 65. 1841; Hook.f., Fl. Brit. India 6: 47. 1890. *A. lineare* Hook.f., Fl. Brit. India 6: 47. 1890.

Syntypes: India: Tamil Nadu, *Wight* in Wall. Cat. no. 7303 & 7313 (K (images!)).

Etymology: The specific epithet refers to the two basal calli at the mouth of the spur.

Epiphytic robust herbs, *ca* 20 cm high. Stem erect. Leaves coriaceous, keeled, sessile, linear or linear-oblong, unequally bilobed, green with purple spots, *ca* 25 × 2.5 cm. Inflorescence axillary, erect, *ca* 28 cm long; racemes laxly many-flowered. Floral bracts pale, triangular, acute, 3-veined, persistent, *ca* 2.5 ×

2 mm. Pedicel with ovary white or whitish-rose, *ca* 1 cm long. Flowers white or whitish-rose. Dorsal sepal ovate-oblong, acute or obtuse, 5-veined, *ca* 7 × 4 mm; lateral sepals broadly ovate or sub-orbicular, acute or obtuse, *ca* 7.5 × 5.5 mm. Petals narrower, obovate, rounded or obtuse, 3-veined, *ca* 6 × 4 mm. Lip sessile, 3-lobed; laterals erect, rounded, *ca* 2.5 × 2 mm; midlobe flat, ovate, *ca* 10 × 5 mm; disc with two basal calli; spur laterally compressed, incurved, cylindric, *ca* 4 mm long. Column white or whitish-rose, oblong-rectangular, *ca* 3 × 1 mm; foot *ca* 3 mm long. Stigma orbicular. Pollinia 2, sub-globose or obovoid, *ca* 1.8 × 1.5 mm. Capsules fusiform, *ca* 2.5 cm long.

Fl. & Fr.: April-July

Habitat: Occasional in semi evergreen forests (above 800 m).

Distribution: ***World***: India and Sri Lanka. ***India:*** Andaman Is., Andhra Pradesh, Karnataka, Kerala, Odisha and Tamil Nadu. ***Eastern Ghats:*** *AP:* Maredumilli and Chintapalli forests; *KA:* BR hills; *OD:* Govindpali forest; *TN:* Kolli hills.

Specimens Examined: *AP*: Maredumilli (EG), *BR* & *KP* 34661 (SKU); Chintapalli (VSKP), *BR* & *MVS* 34678 (SKU). *KA:* BR hills (CRN), *AVNR* 26915 (MH). *OD:* Govindpali ghat (KP), *SM* 2313 & 10 (CAL). *TN*: Kolli hills, Vattaparai (NK), *KMM* 6036, 6038 (RHT).

Doubtfull species

Aerides crispa Lindl., Gen. Sp. Orchid. Pl. 239. 1833.

Note: Reported by Reddy *et al.* (2005) from the Eastern Ghats of Odisha but we have seen no material from this region. Further field studies are required before conforming its distribution in Eastern Ghats.

Anoectochilus Blume

Bijdr. Fl. Ned. Ind. 8: 411. 1825.

World:44/India:7/EG:1

Type: *Anoectochilus setaceus* Blume

Etymology: The generic name is derived from the Greek words, *anoectos* (open) and *cheilos* (lip), refers to the nature of lip.

Distribution: Tropical & Subtropical Asia to Pacific.

Anoectochilus elatus Lindl., J. Proc. Linn. Soc., Bot. 1: 178. 1857; Hook.f., Fl. Brit. India 6: 95. 1890 (*'elatior'*); C.E.C. Fisch., Fl. Madras 1453. 1928; Seidenfaden in K.M. Matthew, Fl. Tamilnadu Carnatic 3: 1557. 1983.

Type: India: Tamil Nadu, Nilgiri hills, Wallaghaut, *Mc.Ivor* 59 (holo: K (image!)).

Etymology: The specific epithet refers to long inflorescence of the species.

Terrestrial herbs, *ca* 30 cm high. Stem creeping at base and erect above, cylindrical, basal part glabrous and hairy above. Leaves 3-5, orbicular-ovate, base rounded, apex acute, velvety dark-green with golden nerves, *ca* 8 × 4 cm; petiole above loose and below sheathing, *ca* 1.5 cm long. Inflorescence slender, *ca* 25 cm long; spike laxly 2-4-flowered (2-12-flowered), glandular pubescent. Floral bracts shorter or longer than the ovary, pale rose, sheathing, ovate-lanceolate, acuminate, glandular pubescent, 1-2-veined, *ca* 1.5 × 0.6 cm. Ovary sub-sessile, greenish-rose, glandular pubescent, *ca* 2 cm long. Sepals whitish-pale inner side, greenish-rose outer side, acuminate, 1-veined, glandular pubescent outer side; dorsal sepal ovate-lanceolate, *ca* 1.2 × 0.5 cm; lateral sepals

Bulbophyllum crassipes Hook.f., Fl. Brit. India 5: 760. 1888; Misra, Orchids Orissa 467. 2004.

Syntypes: India: Sikkim, Terai, Dulkajhar, Clarke 36904 (K (image !)); Martabania, *Wallich* 1990.2 (K); Arracan Hooker (K & CAL); Malaysia: Kerr s.n. (K).

Etymology: The specific epithet refers to the nature of the peduncle.

Epiphytic herbs. Rhizome creeping, woody, branched; pseudobulbs greenish, ribbed or smooth, narrowly ovoid, *ca* 4 × 2 cm. Leaf coriaceous, oblong, obtuse, *ca* 12 × 3.5 cm. Inflorescence erect, cylindrical, *ca* 13 cm long; spike densely many-flowered; peduncle shorter or as long as the pseudobulb. Floral bracts longer than the ovary, yellow, sheathing, oblong-lanceolate, margins feebly undulate, 3-veined, *ca* 7 × 3 mm. Pedicel and ovary pale yellow, *ca* 6 mm long. Flowers arranged in spirals. Sepals and petals yellowish-orange with densely spotted maroon or red; dorsal sepal concave, oblong, apex aristate, 5-veined, warted on outer surface, *ca* 5 × 2.5 mm; lateral sepals cohering their inner edges and forms boat-shaped structure, ovate-lanceolate, 4-veined, warted on outer surface, *ca* 8 × 3.5 mm; petals erect, narrowly triangular, margins sparingly and feebly dentate, apex aristate, 1-veined, *ca* 3.5 mm long, *ca* 2 mm broad at base. Lip mobile, recurved, oblong-ovate, margins feebly crenulated and minutely beaded, *ca* 5 × 3.5 mm; base with two recurved auricles. Column pale yellow, oblong, *ca* 2.2 × 1 mm, winged; stelidia erect, triangular; foot *ca* 2.5 mm long. Stigma oblong-orbicular. Pollinia 4, oblong-ovoid, 0.8 × 0.6 mm. Capsules ellipsoid, *ca* 1.2 cm long.

Fr. & Fr.: October-December.

Habitat: Rare in semi-evergreen and moist deciduous forests (above 500 m).

Distribution: ***World:*** Bhutan, India, Laos, Malaysia, Myanmar, Thailand and Vietnam. ***India:*** Andaman Is., Assam, Jharkhand, Meghalaya, Odisha and Sikkim. ***Eastern Ghats:*** *OD:* Kapilas hills, Rebana forest and Similipal forest.

Specimens Examined: *OD:* Similipal forest-Talabandha (MB), *SM* 87 (CAL); Similipal forest-Podadiha (MB), *Bahali & Agrawala* 429 (CAL); Rebana forest-Panasapani (KJ), *SM* 745 (CAL).

Bulbophyllum fischeri Seidenf., Dansk Bot. Ark. 29: 202. 1973 (1974); Seidenfaden in K.M. Matthew, Fl. Tamilnadu Carnatic 3: 1591. 1983. *Cirrhopetalum gamblei* Hook.f., Fl. Brit. India 5: 778. 1890; C.E.C. Fisch., Fl. Madras 1420. 1928.

Syntypes: India: Tamil Nadu, Nilghiri hills, Conoor, *Gamble* 13168 (CAL & K (images!); Karnataka, Bababoodan hills, *Law s.n.* (K (image!)).

Etymology: This species was named in honour of C.E.C. Fischer (1874-1950), who collected plants from Madras Presidency and one of the author of Flora of the Presidency of Madras.

Epiphytic or lithophytic herbs. Rhizomes creeping, woody, branched; pseudobulbs brownish-red, smooth, globose-ovoid, *ca* 2 × 2 cm. Leaf coriaceous, sessile or sub-sessile, oblong, obtuse, *ca* 8 × 1.5 cm. Inflorescence erect, slender, *ca* 7 cm long; umbels 2-8-flowered. Floral bracts shorter than the ovary, pale, lanceolate, acuminate, 1-veined, *ca* 4 × 1 mm. Pedicel and ovary yellowish-maroon, *ca* 5 mm long. Sepals yellow with base densely striped and mottled with maroon, 5-veined; dorsal broadly ovate, obtuse, *ca* 5 × 3 mm; laterals cohering along the upper margin, widely separated below, falcate, lanceolate or linear-oblong, glandular pubescent, *ca* 15 × 4.5 mm. Petals erect, ovate-oblong or broadly oblong, obtuse to rounded, 3-veined, cream with maroon striations, *ca* 2.5 × 1.5 mm. Lip mobile, yellow with deep maroon striation at base, strongly recurved, ovate-cordate, rounded, *ca* 1.5 × 1 mm. Column pale yellow, oblong, *ca* 4 × 1.2 mm, winged; stelidia erect, triangular; foot *ca* 5 mm long. Stigma oblong-orbicular. Pollinia 4, oblong-ovoid, 0.6 × 0.4 mm.

Fl.: October-November.

Habitat: Rare in densely moss-covered tree trunks and rocks in evergreen forests (above 1000 m).

Distribution: Endemic to South India (Karnataka, Kerala and Tamil Nadu). ***Eastern Ghats:*** *TN:* Kolli hills, Sirumalai hills and Shevaroy hills.

Specimens Examined: *TN:* Kolli hills, Kulivalavu shola (NK), *SKS* 1038 (SGH); Shevaroy hills (SLM), *AVNR* 18248 (MH); Guthriyan hills (DMP), *KMM & NV* 23954, *KMM* 27250 (RHT).

Bulbophyllum guttulatum (Hook.f.) N.P. Balakr., J. Bombay Nat. Hist. Soc. 67: 66. 1970; Misra, Orchids Orissa 474. 2004. *Cirrhopetalum guttulatum* Hook.f., Fl. Brit. India 5: 776. 1896.

Syntypes: Nepal: *Wallich s.n.* (BM); *Wallich* 1984 (BM) and India: Sikkim, *Hooker* 32 (K).

Etymology: The specific epithet refers to sepals and petals with purple spotted nature.

Epiphytic herbs. Rhizome creeping, woody, branched; pseudobulbs angularly ridged, narrowly ovoid. Leaf coriaceous, oblanceolate or oblong, obtuse, *ca* 10 cm long. Inflorescence longer than the leaf, suberect; umbels 3-8-flowered. Floral bracts shorter than the ovary, lanceolate. Pedicel and ovary *ca* 1.5 cm long. Flowers pale cream or yellow with densely spotted with purple. Dorsal sepal broadly ovate, *ca* 6.5 mm long; lateral sepals free, ovate-lanceolate, twisted from base, *ca* 12 mm long. Petals ovate-oblong. Lip strongly deflexed from the middle, narrowly ovate, truncate. Column oblong; stelidia elongated, long-caudate.

Fl.: June-July.

Habitat: Rare in semi-evergreen forest (above 1100 m).

Distribution: ***World:*** Bhutan, India, Nepal and Myanmar ***India:*** Arunachal Pradesh, Nagaland, Meghalaya, Odisha, Sikkim, West Bengal and Uttarakhand. ***Eastern Ghats:*** *OD:* Kutia hills.

Specimens Examined: *OD:* Kutia hills (KP), *SM* 27 (CAL).

Bulbophyllum kaitiense Rchb.f., Ann. Bot. Syst. 6: 262. 1861; Seidenfaden in K.M. Matthew, Fl. Tamilnadu Carnatic 3: 1592. 1983. *Cirrhopetalum nilgherrense* Wight, Icon. Pl. Ind. Orient. 6: t. 1654. 1853; Hook.f., Fl. Brit. India 5: 778. 1888; C.E.C. Fisch., Fl. Madras 1420. 1928.

Syntypes: India: Tamil Nadu, Nilgiri hills, Conoor, Kaitii, *Wight s.n.* & 2957 (K (images!)).

Etymology: This species is named after the type locality name, Kaitie (Nilgiri hills).

Epiphytic or lithophytic herbs. Rhizomes creeping, woody, branched; pseudobulbs light green, smooth or longitudinally ridged, ovoid, *ca* 3 × 1.5 cm. Leaf coriaceous, sessile, oblong or linear-oblong, obtuse or emarginated, *ca* 12 × 2.5 cm. Inflorescence shorter than the leaf, *ca* 10 cm long; umbels 6-8-flowered. Floral bracts green, shorter than the ovary, oblong-ovate, acute, *ca* 2.6 × 1.5 mm. Pedicel and ovary greenish-yellow, *ca* 4 mm long. Flowers greenish-yellow mottled with red, papillose. Dorsal sepal ovate, acute, 5-veined, *ca* 2.6 × 1.4 mm; lateral sepals connate inner margins, linear-lanceolate, *ca* 25 × 3 mm. Petals falcate, broadly oblong, acute, 5-veined, *ca* 1.8 × 1 mm. Lip mobile, fleshy, sessile or sub-sessile, recurved, tongue-shaped, oblong-ovate, acute or obtuse, 3-5-veined, hairy on the back side, *ca* 2 × 1 mm. Column greenish-yellow, oblong,

1.8 mm long; stelidia erect, elongated; foot *ca* 4 mm long, pubescent within. Stigma oblong-triangular. Pollinia 4, oblong-ovoid, 0.8 × 0.8 mm.

Fl.: November-February.

Habitat: Rare on tree trunks and rocks of moist deciduous forest (600-1500 m).

Distribution: Endemic to South India (Kerala and Tamil Nadu). ***Eastern Ghats:*** *AP:* Talkona; *TN:* Kolli hills and Shevarai hills.

Specimens Examined: *AP*: Talkona (CTR), *BSS* & *SKB* 30772 (SKU). *TN*: Kolli hills, Madal shola (NK), *KMM* 63135 (RHT), Kolli hills, Kulivalvau shola (NK), *KMM* 63632 (RHT); Koll hills (NK), *GP* 2006 (RHT); Kolli hills (NK), *PP* 21697 (RHT); Kolli hills (NK), *PP* & *CM* 2008 (RHT); Kolli hills (NK) *KMM* & *SJB* 28266 (RHT); Kolli hills (NK), *KMM* & *VAA* 872 (RHT); Shevarai hills, Yercaud (SLM); *KMM* 30455 (RHT); Shevarai hills, Shervaroyan temple (SLM), *KMM* 23331 (RHT).

Bulbophyllum macraei (Lindl.) Rchb.f., Ann. Bot. Syst. 6: 263. 1861; Misra, Orchids Orissa 471. 2004. *Cirrhopetalum macraei* Lindl., Gen. Sp. Orchid. Pl. 59. 1830; Hook.f., Fl. Brit. India 5: 780. 1888 (non Wight 1851).

Type: Sri Lanka: *Macrae* 8 (holo: K).

Etymology: This species was named in honour of J. Macrae (1800-1830), who collected plants from Sri Lanka (Ceylon) and also type material.

Epiphytic or lithophytic herbs. Rhizome creeping, woody, covered with old sheath-fibres; pseudobulbs narrow-ovoid or globose-ovoid. Leaf oblong or narrowly elliptic, *ca* 17 cm long, petioled. Umbels 2-5-flowered. Floral bracts shorter than the ovary, ovate-lanceolate. Pedicel and ovary *ca* 1 cm long. Sepals light-yellow flushed with maroon; dorsal sepal ovate-lanceolate, acuminate; lateral sepals coherent inner margins, linear-lanceolate, *ca* 3 cm long. Petals light maroon, ovate. Lip bright yellow, ovate-lanceolate. Capsules ellipsoid, *ca* 3 cm long.

Fl. & Fr.: July-October

Habitat: Rare in evergreen and semi evergreen forests (above 800 m).

Distribution: ***World:*** India, Japan, Sri Lanka, Taiwan and Vietnam. ***India:*** Arunachal Pradesh and Odisha. ***Eastern Ghats:*** *OD:* Similipal forest.

Specimens Examined: *OD:* Similipal forest-UBK (MB), *SM* 993 & 1272 (CAL).

Bulbophyllum panigrahianum S. Misra, Nordic J. Bot. 6: 25. 1986. *Cirrhopetalum panigrahianum* (S. Misra) S. Misra, J. Orchid Soc. India 11: 54. 1997; Misra, Orchids Orissa 457. 2004.

Type: India: Odisha, Kendujhar district, Rebana forest, *SM* 698 (holo: CAL!).

Etymology: This species was named in honour of Dr. G. Panigrahi (1924-2004), the renowned plant taxonomist from India.

Epiphytic or lithophytic herbs. Rhizome creeping, woody; pseudobulbs ovoid, *ca* 1 × 1 cm, covered with dark brown foliar sheaths. Leaf coriaceous, oblong-oblanceolate, rounded, *ca* 15 × 3.8 cm; petiole *ca* 2 cm long. Inflorescence erect, *ca* 12 cm long, racemes or umbel, with 4-10-flowered. Floral bracts oblong-lanceolate, acuminate. Pedicel and ovary *ca* 8 mm long. Flowers waxy white flushed with maroon. Sepals and petals ovate-lanceolate, mucronate, 3-7-veined; dorsal sepal cymbiform, *ca* 8 × 4 mm; lateral sepals connate with their distal upper edges, *ca* 10 × 3 mm; petals *ca* 3.6 × 2 mm. Lip recurved, narrowly ligulose, *ca* 2.7 × 1 mm, upper face with papillae. Column winged; stelidia linear. Pollinia 4, globose-ovoid. Capsules ellipsoid, *ca* 1 cm long.

Fl. & Fr.: June-January

Habitat: Rare in evergreen and semi evergreen forests (500-800 m).

Distribution: Endemic to Odisha. ***Eastern Ghats:*** *OD:* Rebana forest and Similipal forest.

Specimen Examined: OD: Rebana forest (KJ), *SM* 698 (CAL).

Bulbophyllum polyrrhizum Lindl., Gen. Sp. Orchid. Pl. 53. 1830; Hook.f., Fl. Brit. India 5: 767. 1888; Misra, Orchids Orissa 464. 2004.

Type: India: Sikkim, *Wallich* icon.

Etymology: The specific epithet refers to the nature of the rhizomes.

Epiphytic or lithophytic herbs. Rhizomes creeping, slender, branched; pseudobulbs crowded, greenish-purple, globose-ovoid or oblong-ovate, *ca* 1 × 0.5 cm. Leaf linear-oblong, acute, sessile, *ca* 9 × 0.8 cm. Inflorescence in leafless condition, erect, *ca* 6 cm long; raceme laxly few-flowered. Floral bracts shorter than the ovary, yellow, oblong-ovate, acute, *ca* 1.5 × 1 mm. Pedicel and ovary greenish-yellow, *ca* 3 mm long. Flowers yellow. Sepals ovate-oblong, acute, 3-veined; dorsal *ca* 4 × 2.5 mm; lateral sepals *ca* 5 × 2 mm. Petals erect, lanceolate, acute, *ca* 2.5 × 1.2 mm. Lip recurved, stipitate, oblong-ovate, rounded, crenulated, 3-veined, scabrous, *ca* 3.5 mm long. Column yellow, ovate-oblong, *ca* 2 × 1.2 mm; stelidia toothed; foot *ca* 2 mm long. Stigma sub-quadrangular. Pollinia 4, sub-ovoid, 0.4 × 0.2 mm. Capsules ellipsoid, *ca* 5 mm long.

Fl. & Fr.: March-October.

Habitat: Occasional on moss covered tree trunks or rocks of semi evergreen and moist deciduous forests (600-1000 m).

Distribution: ***World:*** India, Nepal, Myanmar and Thailand. ***India:*** Arunachal Pradesh, Meghalaya, Mizoram, Odisha, Sikkim and Uttarakhand. ***Eastern Ghats:*** *OD:* Mahendragiri hills and Similipal forest.

Specimens Examined: *OD:* Similipal forest-Meghasani hills (MB), *SM* 1123 (CAL); Singaraja hills (GJP), *SM* 198 (CAL); Panechpatmali hills (KP), *SM* 236 (CAL).

Bulbophyllum sterile (Lam.) Suresh, Interpret. Van Rheede's Hort. Malab. 298. 1988. *Epidendrum sterile* Lam., Encycl. 1: 189. 1783. *Bulbophyllum nilgherrense* Wight, Icon. Pl. Ind. Orient. 5: t. 1650. 1851; Hook.f., Fl. Brit. India 5: 761. 1888; C.E.C. Fisch., Fl. Madras 1418. 1928; Seidenfaden in K.M. Matthew, Fl. Tamilnadu Carnatic 3: 1591. 1983.

Type: India: Malabar, Rheed, Hort. Malab. (theka-maravara) 12: t. 22. 1693.

Etymology: The specific epithet refers to the Rheede figure without flowering.

Epiphytic or lithophytic herbs. Rhizome creeping, woody, branched; pseudobulbs yellowish-green, ridged or smooth, conical-ovoid or globular-conical, *ca* 3 × 1.5 cm. Leaf fleshy, sessile, narrowly elliptic-oblong, obtuse, *ca* 10 × 2.5 cm. Inflorescence drooping, *ca* 9 cm long; raceme densely many-flowered. Floral bracts as long as or longer than the ovary, brownish-green, ovate-oblong, acute, 3-veined, *ca* 5 × 3 mm. Pedicel and ovary greenish-maroon, *ca* 4 mm long. Flowers fragrant, greenish-yellow with pink dots. Dorsal sepal concave, broadly ovate, acute or obtuse, 3-veined, *ca* 4.5 × 2.8 mm; lateral sepals inner margins connate and outer margins upcurved to form boat-shaped structure, oblong-lanceolate, *ca* 8 mm long, *ca* 4 mm broad at base. Petals erect, triangular-ovate, margins shallowly dentate or undulate, apex awned, 1-veined, *ca* 2.8 mm long, 2 mm broad at base. Lip deflexed, tongue-shaped, obtuse, entire or shallowly 3-lobed, with basal auricles, *ca* 5 × 2 mm. Column yellowish-green, oblong, 3 × 1.4 mm, winged; stelidia erect; foot *ca* 3 mm long. Stigma quadrate-orbicular. Pollinia 4, oblong-obovoid, 1 × 0.8 mm.

Fl.: November-February.

Habitat: Rare in semi-evergreen and moist deciduous forests (600-1000 m).

Distribution: ***World:*** Bangladesh and India. ***India:*** Andhra Pradesh, Karnataka, Kerala, Maharashtra and Tamil Nadu. ***Eastern Ghats:*** *AP:* Lankapakulu and Seshachalam hills; *TN:* Pachamalai, Kolli hills and Shevaroy hills.

Specimens Examined: *AP:* Lankapakulu MPCA (VSKP), *K.N. Reddy & CSR* 1754 (FRLH). *TN:* Kolli hills, Kulivalavu shola (NK), *KMM* 62756 (RHT); Kolli hills, Arapaleshwarar top falls (NK), *KMM & CR* 2103, *KMM* 3289, *KMM & VKM*, 9670 (RHT); Shevarai hills, Shervarayan temple (SLM), *KMM* 62049 (RHT); Shevarai hills, Shervarayan temple-Yercaud (SLM), *SP* 70148 (RHT); Pachamalai (TRP), *KMM* 25966 (RHT).

Bulbophyllum tremulum Wight, Ic. Pl. Ind. Or. 5(1): 20. t. 1749. 1841; Hook.f., Fl. Brit. India 5: 763. 1890; C.E.C. Fisch., Fl. Madras 1418.1928; Kottaimuthu & Ponnuchamy, Bio Diver. 42: 6203. 2012.

Type: India: Kerala, Wynaud, *Jerdon* and *Cotton s.n.* (not located).

Etymology: The specific epithet refers to long hairs on the back of the lip, perpetually in a state of tremulous motion.

Epiphytic herbs. Rhizomes creeping, woody; pseudobulbs depressed, rigid, ovoid-conical, *ca* 1.7 × 1.5 cm. Leaf oblong-lanceolate or elliptic-oblong, obtuse-acute, *ca* 7 × 2 cm. Racemes much longer than the leaf, laxly few-flowered. Flowers purplish, 7.5-15 cm long. Dorsal sepal ovate-oblong; lateral sepals lanceolate, ciliate. Petals scaly, fringed with ciliate hairs. Lip linear-oblong, fringed with long acicular hairs. Capsules ellipsoid, *ca* 2 cm long.

Fl. & Fr.: March-August

Habitat: Rare in semi-evergreen forests (above 600 m).

Distribution: Endemic to Karnataka, Kerala and Tamil Nadu. ***Eastern Ghats:*** *TN:* Sirumalai hills.

Specimens Examined: *TN*: Sirumalai hills, Nerkuthu sholai, *R. Kottaimuthu* 30400 (ATREE) & Ponnuruki sholai, *R. Kottaimuthu* 30678 (ATREE).

Bulbophyllum triste Rchb.f., Ann. Bot. Syst. 6: 253. 1861; Hook.f., Fl. Brit. India 5: 768. 1888; Haines, Bot. Bihar & Orissa, 1176. 1924; Misra, Orchids Orissa 484. 2004.

Types: India: Meghalaya, Khasia hills, *Lobb s.n.*(K & W (images!)).

Etymology: The specific epithet refers to the dark-coloured flowers of the species.

Epiphytic herbs. Rhizome creeping, woody; pseudobulbs crowded, conico-globose. Leaves 2, membranous, sessile, oblong-lanceolate, *ca* 13 cm long. Inflorescence drooping, *ca* 10 cm long; spike densely many-flowered. Floral bracts yellow, ovate or triangular. Pedicel and ovary *ca* 3 mm long. Dorsal sepal narrowly ovate, *ca* 3.8 mm long; lateral sepals cohering at tips, gibbous near base, oblong-lanceolate, margin strongly recurved, *ca* 4.5 mm long. Petals erect,

Habitat: Rare, it grows near the perennial streams under deep shady condition in humus rich soils in evergreen forests (above 1000 m).

Distribution: ***World*:** Africa, Bhutan, Borneo, China, India, Japan, Java, Madagascar and the Mascarenes, Myanmar, Nepal, Sri Lanka, Sumatra, Thailand and Vietnam. ***India:*** Arunachal Pradesh, Karnataka, Kerala, Manipur, Meghalaya, Mizoram, Nagaland, Sikkim and Tamil Nadu. ***Eastern Ghats:*** *TN:* Kolli hills.

Specimens Examined: *TN:* Kolli hills, Peria solai (NK), *AA. Ansari & P. Dwarakan* 78825 (MH).

Calanthe triplicata (Willemet) Ames, J. Sci., C 2: 326. 1907; Misra, Orchids Orissa 506. 2004; Seidenfaden in K.M. Matthew, Fl. Tamilnadu Carnatic 3: 1573. 1983. *Orchis triplicata* Willemet, Ann. Bot. (Usteri) 18: 52. 1796. *Calanthe veratrifolia* R.Br. ex Ker Gawl., Bot. Reg. 9: t. 720. 1823; Hook.f., Fl. Brit. India 5: 851. 1888; C.E.C. Fisch., Fl. Madras 1432. 1928.

Type: Indonesia: Rumphius, *Herb. Amboin.* 6: t. 52. 1750.

Etymology: The specific epithet refers to the strongly 3-lobed lip.

Terrestrial herbs, *ca* 1 m high. Rhizome creeping, woody; pseudobulbs crowded, green, ovoid or conical, *ca* 3 × 1.5 cm. Leaves 3-5, elliptic-oblong or elliptic-ovate, acuminate, pubescent below, *ca* 36 × 12 cm; petiole grooved, sheathing, *ca* 16 cm long. Inflorescence sparsely pubescent, *ca* 1 m long; raceme densely many-flowered. Floral bracts shorter than the ovary, green, ovate-lanceolate, acute, pubescent, *ca* 1.2 × 5 cm. Pedicel with ovary white, *ca* 3.5 cm long, pubescent. Flowers white with red-orange callus. Sepals broadly ovate or elliptic-obovate, acuminate, pubescent outer surface, 3-5-veined, *ca* 1.5 × 0.7 cm. Petals oblanceolate, acuminate, narrowed at base, pubescent outer surface, 3-veined, *ca* 1.4 × 0.6 cm. Lip sessile, 3-lobed, *ca* 1.5 cm long; laterals spreading, oblong, rounded; mid-lobe deeply bilobuled; lobules diverging, recurved, falcately oblong; disc tubercled; spur cylindrical, *ca* 1.8 cm long. Column white, truncate, *ca* 6 × 4 mm, pubescent. Pollinia 8, club-shaped, 1.7 × 0.6 mm. Capsules ellipsoid, *ca* 4 cm long.

Fl. & Fr.: June-October.

Habitat: Rare, it grows near the perennial streams under deep shady condition in humus-rich soils in evergreen forests (above 800 m).

Distribution: ***World:*** Australia, Bhutan, Borneo, Cambodia, China, Fiji, Fujian, India, Japan, Java, Laos, Lesser Sunda Is., Madagascar, Malaysia, Moluccas, Myanmar, New Caledonia, New Guinea, Pacific Is., Philippines, Seychelles, Sri Lanka, Sulawesi, Sumatra, Thailand, Taiwan and Vietnam. ***India:*** Andaman Is., Arunachal Pradesh, *Assam,* Karnataka, Kerala, Mizoram, Nagaland, Odisha, Sikkim and Tamil Nadu. ***Eastern Ghats:*** *OD:* Similipal forest; *TN:* Kolli hills.

Key to the species

1a. Leaves pale green; lip white with a triangular green patch on central part near mesochile...**C. flabellata**

1b. Leaves brownish-green to reddish-brown with pale yellow tinge; lip white with two pale green spots near the base of epichile.................................**C. parvifolia**

Cheirostylis flabellata (A. Rich.) Wight, Icon. Pl. Ind. Orient. v. 16. t. 1727 (*Monochilus*). 1851; C.E.C. Fisch., Fl. Madras 1454.1928; *Goodyera flabellata* A. Rich., Ann. Sci. Nat., Bot. ser. 2, 15: 79. 1841. *Monochilus flabellatus* (A. Rich.) Wight, Icon. Pl. Ind. Orient. t. 1727. 1851.

Types: India: Indes Orientales, Nillighery, Perrottet *s.n.* (P & L).

Etymology: The specific epithet refers to the shape of the lip.

Terrestrial herbs; rhizomes *ca* 10 cm long, decumbent, internodes 1-1.4 cm long. Leaves 3-4, pale green, ovate, *ca* 3 × 1.7 cm. Inflorescence *ca* 15 cm long; raceme 5-10-flowered; peduncle hairy at base and glabrous apically. Floral bracts lanceolate, glabrous. Ovary sessile, glabrous, *ca* 8 mm long. Sepals light greenish-brown, connate to form sepaline tube, 4.5-5.5 mm long, glabrous. Petals white, falcate, *ca* 5 mm long. Lip connate with column at base; hypochile saccate, keeled at bottom, with 2-3 horn-like papillae within each half of sac; mesochile short, claw-like; epichile white, with a triangular green patch on central part near mesochile, dilated and bilobed, orbicular in outline, recurved at apex; lobules semiorbicular, deeply with 7-9-laciniate. Stylids 2, as long as and parallel to rostellar arms. Stigma lobes 2; rostellum bifid. Pollinia 2, clavate.

Fl.: August-December.

Habitat: Occasional, it grows under shade condition in humus rich soils of semi evergreen and moist deciduous forests (above 1000 m).

Distribution: ***World:*** Bhutan, China, India, Myanmar, Sri Lanka, Taiwan and Vietnam. ***India:*** Almost common. ***Eastern Ghats:*** *TN:* Kolli hills and Shevaroy hills.

Specimen Examined: *TN:* Shevaroy hills, Yercaud (SLM), *SKS* 824 (SGH); Shevaroy hills, Sanyasimalai R.F. (SLM), *S. Kaliamoorthy & T.S. Saravanan s.n.* (BSI-SRC, Yercaud).

Cheirostylis parvifolia Lindl., Edward's Bot. Reg. 25 (Misc.): 19. 1839; Hook.f., Fl. Brit. India 6: 105. 1890; Dash *et al.,* J. Threat. Taxa 7(1): 6827. 2015.

Type: Sri Lanka (Ceylon): *Loddiges s.n.* (holo: K (image!)).

Etymology: The specific epithet refers to small leaves of the species.

Terrestrial herbs, *ca* 18 cm high. Roots tufted, fibrous. Rhizome creeping, 2-7 cm long, brownish-green. Stem brownish-green, *ca* 7 cm long. Leaves 2-8, withered during the flowers, ovate to ovate-lanceolate, acute or acuminate, *ca* 3.5 × 2.5 cm, brownish-green to reddish-brown with pale yellow tinge; petiole sheathing, *ca* 1 mm long. Inflorescence covered with glandular pubescence; racemes sub-densely 2-many-flowered. Floral bracts shorter than the ovary, light pink, ovate, acuminate, sparsely pubescent, 1-veined, *ca* 8 × 2.5 mm. Pedicel with ovary pale green, *ca* 2.1 cm long. Sepals connate to form sepaline tube, pale green to greenish-white with pinkish tinge at apex, pale pink during maturity, glabrous to laxly glandular pubescent outer surface near the base; dorsal lobe broadly ovate to ovate-lanceolate, *ca* 2.8 × 1.3 mm; lateral lobes lanceolate, *ca* 2.8 × 1 mm. Petals white, falcate, oblong-spathulate to narrowly obovate, obtuse, glabrous, 1-veined, *ca* 3 × 1 mm. Lip white with two pale green spots near the base of epichile, *ca* 4 mm long; hypochile saccate, *ca* 0.6 × 0.7 mm, with 2-4 papillose calli inside on each side; mesochile linear, *ca* 1 × 0.7 mm; epichile semicircular, *ca* 1.5 × 2 mm, bilobed; lobules with minute hairs near the base. Stylidia 2, as long as or some time shorter than the rostellar arms. Stigma lobes 2, separated; rostellum deeply bifid. Pollinia 2, ovate-oblong to pyriform, 1 mm long. Capsules obconical.

Fl. & Fr.: August-January.

Habitat: Rare, it grows under shade condition in humus rich soils of semi evergreen and moist deciduous forests (700-1000 m).

Distribution: ***World:*** India and Sri Lanka. ***India:*** Kerala, Maharashtra and Odisha. ***Eastern Ghats:*** *OD:* Mahendragiri hills.

Specimens Examined: *OD:* Mahendragiri hills (GJM), *Prasad* 9959 (RPRC).

Chiloschista Lindl.

Edwards's Bot. Reg. 18: t. 1522. 1832.

World: 20/India:4/EG:1

Type: *Chiloschista usneoides* (D. Don) Lindl.

Etymology: The generic name is derived from Greek words, *cheilos* (lip) and *chistos* (cleft) refers to the shape of the lip.

Distribution: Tropical and Subtropical Asia to West Pacific.

Chiloschista parishii Seidenf., Opera Bot. 95: 176. 1988; Misra, Orchids Orissa 615. 2004. *Sarcochilus luniferus* sensu Hook.f., Fl. Brit. India 6: 37. 1890.

Type: Myanmar (Burma): Moulmein, *Parish* 55 (holo:W).

Etymology: The species was named in honour of C.S.P. Parish (1822-1897), who collected plants from Myanmar (Burma) and Andaman Islands and the type specimens.

Epiphytic herbs, without stem and leaves. Roots tufted, green, spreading, cylindrical, unbranched. Inflorescence arising from the base of the root stocks, spreading or sub-erect, pubescent or hirsute, *ca* 20 cm long; raceme laxly few-many-flowered. Floral bracts shorter than the ovary, pale white, membranous, ovate-triangular, *ca* 2.5 × 2 mm. Pedicel with ovary *ca* 1 cm long, pubescent. Flowers scented, fleshy. Sepals and petals greenish-yellow with brown blotched, hirsute on outer surface, apex rounded; sepals similar, *ca* 5 × 4 mm; dorsal sepal incurved, broadly ovate; lateral sepals ovate-orbicular. Petals broadly ovate-sub-quadrate, *ca* 4.5 × 4 mm. Lip white with brownish veins, strongly inflexed, sessile, 3-lobed; lobes forming pouch shape; laterals oblong, *ca* 3 × 2 mm; midlobe inconspicuous; disc with hairy callus. Column greenish-yellow, subquadrate-ovate, 2 × 1 mm; foot incurved, *ca* 2.8 mm long, notched at apex. Stigma broadly ovate. Anther with two lateral filiform setae; pollinia 4, sub-globose, 1 × 0.9 mm. Capsules fusiform, *ca* 3 cm long.

Fl. & Fr.: March-July.

Habitat: Rare, it grows on fully moss covered tree trunks in semi evergreen and moist deciduous forests (500-1000 m).

Distribution: ***World:*** Bhutan, China, India, Myanmar, Thailand and Vietnam. ***India:*** Andaman Is., Odisha, Sikkim and West Bengal. ***Eastern Ghats:*** *OD:* Rebana forest and Similipal forest.

Specimens Examined: *OD:* Similipal forest-UBK (MB), *SM* 810 (CAL).

Chrysoglossum Blume

Bijdr. Fl. Ned. Ind. 337. 1825.

World:4/India:2/EG:1

Type: *Chrysoglossum ornatum* Blume

Etymology: The generic name is derived from Greek words, *chryso* (golden) and *glossa* (tongue), refers to the flower colour and shape.

Distribution: Tropical and Subtropical Asia to SW. Pacific.

Chrysoglossum ornatum Blume, Bijdr. 338. 1825. *C. maculatum* (Thwaites) Hook.f., Fl. Brit. India 5: 784. 1890; C.E.C. Fisch., Fl. Madras 1421. 1928; Seidenfaden in K.M. Matthew, Fl. Tamilnadu Carnatic 3: 1603. 1983; Misra, Orchids India 287. 2007. *C. erraticum* Hook. f., Fl. Brit. India 5: 784. 1890.

Type: Java: Salak, Tjapus River, *Blume s.n.* (holo:L).

Etymology: The specific epithet refers to the lip lamellae of the species.

Habitat: Rare, it grows in moist deciduous forests (above 600 m).

Distribution: ***World*:** India, Sri Lanka and Thailand. ***India:*** Andhra Pradesh, Karnataka, Kerala, Tamil Nadu and West Bengal. ***Eastern Ghats:*** *AP:* Seshachalam hills.

Specimens Examined: *AP:* Tirumala hills (CTR), *DR* 263 & 849 (SVU).

Coelogyne Lindl.

Coll. Bot. t. 33. 1824.

World:207/India:42/EG:2

Type: *Coelogyne nitida* Lindl.

Etymology: The generic name is derived from Greek words, *coelo* (hollow) and *gynus* (stigma), refers to the deep hollow stigma.

Distribution: Tropical and Subtropical Asia to West Pacific.

Key to the species

1a. Lip white with two small yellow patched; disc with 2 lamellae .... **C. breviscapa**

1b. Lip whitish-yellow below the midlle; disc with 3 lamellae .... **C. nervosa**

Coelogyne breviscapa Lindl., Fol. Orchid. 5: 4. 1854; Hook.f., Fl. Brit. India 5: 91. 1890; C.E.C. Fisch., Fl. Madras 1430. 1928; Seidenfaden in K.M. Matthew, Fl. Tamilnadu Carnatic 3: 1574. 1983.

Type: Sri Lanka: *Walker s.n.* (K (image!)).

Etymology: The specific epithet refers to the short scape.

Epiphytic or lithophytic herbs. Rhizome horizontal, creeping, woody; pseudobulbs crowded, narrowly oblong or conical, covered with sheathing scales, *ca* 4 × 1.5 cm. Leaves 2, coriaceous, keeled, linear or linear-oblong or liner-elliptic, acute, *ca* 10 × 1.4 cm; petiole *ca* 5 mm long. Inflorescence terminal or arising from the base of pseudobulb, erect, *ca* 11 cm long; raceme laxly 4-7-flowered, peduncle covered closely imbricating sheaths. Floral bracts as long as or longer than the ovary, brown, ovate-lanceolate, acute, *ca* 2.5 × 1 cm. Pedicel with ovary pale white, *ca* 1.5 cm. Flowers scented, white. Sepals acute, 5-6-veined; dorsal oblong-lanceolate, *ca* 1.3 × 0.5 cm; laterals ovate-oblong, *ca* 1.3 × 0.4 cm. Petals linear, obtuse, 3-veined, *ca* 1.2 × 0.2 cm. Lip sessile, *ca* 1.2 cm long, 3-lobed; laterals embracing the column, auricular; midlobe broadly obovate or orbicular-ovate, apiculate or retuse at apex; disc with 2 erect crisped, undulate lamellae on either side of the median vein. Column white, oblong, winged, *ca* 7 × 2.5 mm. Stigma orbicular. Pollinia 4, pyriform, *ca* 2.5 mm. Capsules ellipsoid, *ca* 3 cm long.

Fl. & Fr.: April-January.

Habitat: Rare, it grows on moss-covered tree trunks and rocks of evergreen forests (above 1200 m).

Distribution: ***World:*** India & Sri Lanka. ***India:*** Karnataka, Kerala and Tamil Nadu. ***Eastern Ghats:*** *KA:* BR hills; *TN:* Kolli hills.

Specimens Examined: *TN:* Kolli hills, Solakkadu (NK), *SKS* 1181 (SGH); Kolli hills (NK), *KMM* 2242 (RHT); Kolli hills, Solakkadu (NK), *KMM* & *VKN* 9666 (RHT).

Coelogyne nervosa A. Rich., Ann. Sci. Nat., Bot. II, 15: 16. 1841; C.E.C. Fisch., Fl. Pres. Madras 1430. 1928.

Types: India: Tamil Nadu, Nilgiri hills, *Richard s.n.* (K & P (images!)).

Etymology: The specific epithet refers to the prominent veined leaves.

Epiphytic or lithophytic herbs. Pseudobulls broadly ovoid, *ca* 3.5 cm, wrinkled. Leaves 2, elliptic or oblong-lanceolate, *ca* 16 cm long, prominently vined. Raceme laxly 2-5-flowered at apex. Floral bracts longer than the ovary, ovate-lanceolate. Pedicel with ovary *ca* 1.5 cm long. Flowers white. Dorsal sepal elliptic-oblong, acute, *ca* 2.6 mm long, 9-many veined; lateral sepals elliptic or elliptic-oblong. Petals elliptic, acute, 3-5-veined. Lip *ca* 2.4 cm long, 3-lobed; laterals embracing the column, yellowish-white; midlobe broadly triangular or ovate, margin obscurely crenulate; disc with 3 longitudinal, erect, crisped lamellae. Column winged, apex hooded. Pollinia 4, pyriform, *ca* 2.5 mm. Capsules ellipsoid.

Fl. & Fr.: June-October.

Habitat: Rare, it grows on moss-covered tree trunks and rocks of evergreen forests (above 800 m).

Distribution: ***World:*** India & Myanmar. ***India:*** Karnataka, Kerala and Tamil Nadu. ***Eastern Ghats:*** *KA:* BR hills.

Specimens Examined: *KA:* Biligiri Rangana hills (CRN), *E. Barnes s.n.* (MH).

Conchidium Griff.

Not. Pl. Asiat. 3: 321. 1851.

World:17/India:5/EG:3

Type: *Conchidium pusillum* Griff.

Etymology: The generic name is derived from Greek word, *konche* (mussel shell) referring to the shell-like appearance of the depressed pseudobulbs.

Distribution: South India to Sri Lanka, Himalaya to temperate East Asia.

Key to the species

1a. Flowers solitary .... **C. reticosum**

1b. Flowers in racemes 2

2a. Flowers pale creamy yellow; lip margins entire .... **C. exile**

2b. Flowers greenish-yellow or pale white; lip margins crenulated .... **C. nanum**

Neotypes: Java: *Lobb s.n.* & 162 (K; designated by Rasmussen).

Etymology: The specific epithet refers to the white hellebore (*Veratrum album* L.) which has similar leaves.

Terrestrial herbs. Stem *ca* 1.2 m high. Leaves elliptic-lanceolate, caudate-acuminate, nearly sessile on the sheaths, *ca* 46 × 10 cm. Panicles *ca* 15 cm long. Flowers *ca* 3.2 cm long, greenish-white. Sepals and petals linear-oblanceolate. Lip linear with the end expanded into an orbicular, apiculate. Pollinia 2.

Fl.: July-August.

Habitat: Rare, it grows along perennial streams of moist deciduous forest.

Distribution: ***World:*** Australia, Borneo, China, Fiji, India, Indonesia, Java, Laos, Malaysia, Moluccas, New Guinea, Pacific Is., Philippines, Ryukyu Is., Samoa Is., Singapore, Solomon Is., Sri Lanka, Sumatra, Thailand, Tonga, Vanuatu and Vietnam. ***India:*** Andaman & Nicobar Is., Andhra Pradesh, Arunachal Pradesh, Assam, Kerala, Meghalaya, Sikkim and Tamil Nadu. ***Eastern Ghats:*** *AP:* Seshachalam hills.

Note: Reported by Fischer (1928) based on Beddome *l.c.* but our subsequent visit, we couldn't locate this species in Kadapa district of Seshachalam hills.

Cottonia Wight

Icon. Pl. Ind. Orient. 5(1): 21. 1851.

Monotypic genus.

Type: *Cottonia macrostachya* Wight

Etymology: The genus was named in honour of Major General Cotton (1792–1874), a British Army officer of the 19th century and who collected plants in South-Western India.

Distribution: India and Sri Lanka.

Cottonia peduncularis (Lindl.) Rchb.f., Cat. Orch.-Samml. Schiller ed. 3: 22. 1857; Misra, Orchids Orissa 572. 2004; Misra *et al.,* Bull. Bot. Surv. India 50: 135. 2008. *C. macrostachya* Wight, Icon. Pl. Ind. Orient. t. 1755. 1851; Hook.f., Fl. Brit. India 6: 26. 1890; C.E.C. Fisch., Fl. Madras 1439. 1928. *Vanda peduncularis* Lindl., Gen. Sp. Orchid. Pl. 216. 1833.

Type: Sri Lanka: *Macrae* 40 (holo:K (image!)).

Etymology: The specific epithet refers to long peduncle of the species.

Epiphytic herbs. Stem erect, stout, *ca* 20 cm long, fully covered with sheathing leaf bases. Leaves coriaceous, sessile, lorate or oblong, apex unequally bilobed, *ca* 15 × 1.5 cm. Inflorescence leaf-opposed, much longer than the leaves, *ca* 1.3 m long; panicle laxly many-flowered. Floral bracts brownish-

obtuse, *ca* 1 × 0.6 cm; disc with 2 yellow keels. Column yellow with maroon stained, oblong-obovate, *ca* 1 × 0.4 cm, winged, with foot. Stigma sub-triangular. Pollinia 2, ovoid, 3 × 1.8 mm. Capsules pyriform, *ca* 6 cm long.

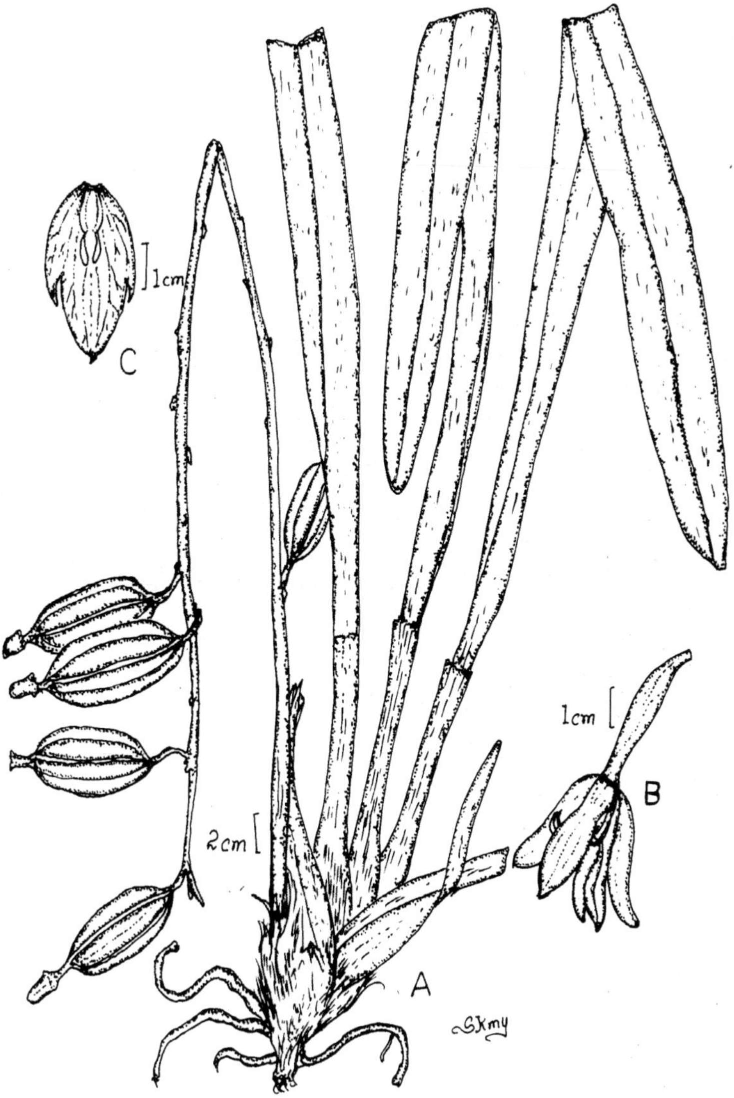

Figure 2. **Cymbidium aloifolim** (L.) Sw.
A. Habit; B. Flower; C. Lip

Fl. & Fr.: March-August.

Habitat: Common in semi evergreen and moist deciduous forests (above 200 m).

Distribution: ***World*:** Bangladesh, Bhutan, Borneo, Cambodia, China, India, Indonesia, Java, Laos, Malaysia, Myanmar, Nepal, Philippines, Sri Lanka, Sumatra, Taiwan, Thailand and Vietnam. ***India:*** Andaman & Nicobar Is., Andhra Pradesh, Arunachal Pradesh, Assam, Chhattisgarh, Goa, Jharkhand, Karnataka, Kerala, Maharashtra, Manipur, Meghalaya, Mizoram, Nagaland, Odisha, Sikkim, Tamil Nadu, Tripura and West Bengal. **Eastern Ghats:** Occasionally found in Eastern Ghats.

Specimens Examined: *AP:* Krishnapuram (CTR), *MVS, VSR* & *KP* 31065 (SKU); Maredumilli (EG), *BR* & *KNR* 10617 (SKU), *BR* & *KP* 34655 (SKU), *GVS* 24250 (MH), *VNS* 340 (CAL); Palagondi (EG), *SS* 5489; Vatangi (EG), *CAB* 12672 (MH); Peddaveedi, Tadepalli (EG), *NRR* 76820 (BSID); Kuntada RF, Maredumilli (EG), *MM* 100731 (MH); Jamparakota (VSKP), *KCJ* 17244 (MH); Near Borra-Anatagiri (VSKP), *NPBK* 801 (CAL); Manda (VZN), *MV* 3775; Near Gummada (SKLM), *GVS* 62431 (CAL). *OD:* Similipal forest-Bania-basa (MB), *Bahali* & *Agrawala* 593 (CAL); Govindpali ghat (KP), *SM* 31 (CAL). *TN:* Javvadhu hills (TRP), *R. Vijayasankar* 43523, 43626 (FRLH); Top Sengatupatty, Pachamalai (TRP), *KMM* 15948 (RHT); Pachamalai (TRP), *DIA* 6969 (RHT); Kolli hills, Arapaleshwarar temple (NK), *KMM* & *CR* 2102, *PP* 21702 & 22498, *NV* 15822 (RHT); Vattaparai, above Puliancholai (NK), *KMM* 5684 (RHT); Solavandhi river (NK), *AM* 13066 (RHT); Periyakalrayan (SLM), *KMM* & *KM* 15210 (RHT), *NV* 22216 (RHT); Shevaroyan, *NV* 22343 (RHT); Chinnakalyaran (SLM), *TSJ* 21988 (RHT); Bodai malai (SLM), *PP* & *CM* 22018 (RHT).

Cymbidium bicolor Lindl., Gen. Sp. Orchid. Pl. 164. 1833; Hook.f., Fl. Brit. India 6: 11. 1890; Misra, Orchids Orissa 526. 2004.

Type: Sri Lanka: *Macrae* 54 (holo: K (image!)).

Etymology: The specific epithet refers to the bicolour flowers of the species.

Epiphytic herbs. Roots copious with needle like rootlets. Stems tufted, pendulous; pseudobulbs compressed, narrowly ovate-conical, nodded, completely covered with leaf-sheaths. Leaves curved, coriaceous, keeled, sessile, loriform, unequally bilobed, *ca* 60 cm long. Inflorescence pendulous, *ca* 35 cm long; raceme laxly many-flowered. Floral bracts maroon, triangular, obtuse. Pedicel with ovary maroon, *ca* 2.5 cm long. Flowers fragrant, maroon with pale yellow margins, fleshy. Sepals oblong or oblong-oblanceolate, rounded. Petals elliptic-oblong or obovate-oblong, rounded. Lip sessile, 3-lobed; laterals shorter than the column, narrowly triangular; midlobe recurved, narrowly ovate-oblong; lamellae sigmoid, clavate. Column oblong, with foot. Pollinia 2, ovoid. Capsules ellipsoid, *ca* 5 cm long.

Fl. & Fr.: March-November

Habitat: Rare in semi evergreen forest (above 500 m).

Distribution: ***World:*** Bangladesh, Borneo, Cambodia, China, Java, India, Laos, Malaysia, Nepal, Philippines, Sri Lanka, Sulawesi, Sumatra, Thailand and Vietnam. ***India:*** Arunachal Pradesh, Assam, Karnataka, Kerala, Odisha, Sikkim, Tamil Nadu, Uttarakhand and West Bengal. ***Eastern Ghats:*** *OD:* Similipal forest.

Specimens Examined: *OD:* Similipal forest-UBK (MB), *SM* 1019 (CAL); Similipal forest-Kabataghai (MB), *SM* 1039 (CAL).

Dendrobium Sw.

Nova Acta Regiae Soc. Sci. Upsal. 6: 82. 1799 (*nom. cons.*).

World:1400/India:116/EG:19

Type: *Dendrobium moniliforme* (L.) Sw. (*typ. cons.*).

Etymology: The generic name is derived from Greek word, *dendron* (tree) and *bios* (life), refers to the epiphytic habit of the genus.

Distribution: Tropical and Subtropical Asia to Pacific and New Zealand.

Key to the species

1a. Stem much branched .... **D. herbaceum**

1b. Stem not branched 2

2a. Flowers in racemes 3

2b. Flowers solitary or in fascicles or cymes 10

3a. Stem reduced to globular or ovoid pseudobulb 4

3b. Stem elongated 7

4a. Racemes terminal on leaf less pseudobulbs .... **D. microbulbon**

4b. Racemes with leafy pseudobulbs 5

5a. Flowers white with pink tinge .... **D. anamalayanum**

5b. Flowers not as above 6

6a. Pseudobulbs conspicuous; lip midlobe margins strongly crisped; disc with cushion-like rugose .... **D. peguanum**

6b. Pseudobulbs inconspicuous; lip midlobe pectinate; disc with 3 broad depressed fleshy truncate ridges .... **D. nanum**

7a. Column foot biloculate; flowers pale yellow with purple or red speckled .... **D. bicameratum**

7b. Column foot not as above; flowers various coloured 8

8a. Flowers magenta with dark veins and transversely markings .... **D. regium**

8b. Flowers not as above 9

9a. Lip yellow-ochre with 2 brownish-purple spots inside, slipper-shaped or open-mouthed pouch .... **D. moschatum**

9b. Lip white with green tinge at base and pink tinge near the apex or not, not slipper-shaped .... **D. wightii**

10a. Flowers solitary .... **D. macraei**

10b. Flowers not as above 11

11a. Flowers in leaf less condition 12

11b. Flowers in leafy condition 16

12a. Lip white with large, yellow central patch; margins glandular-papillose .... **D. crepidatum**

12b. Lip not as above 13

13a. Lip with large purple or red patches 14

13b. Lip without purple patch 15

14a. Petals ovate-lanceolate; disc of lip pubescent .... **D. heterocarpum**

14b. Petals broadly ovate; disc of lip glabrous .... **D. transparens**

15a. Flowers pale-green and turning creamish-yellow or lemon-green at age .... **D. macrostachyum**

15b. Flowers whitish-rose .... **D. aphyllum**

16a. Flowers golden yellow with purple marked on lip .... **D. cathcartii**

16b. Flowers not as above 17

17a. Disc of lip without yellow patch... **D. nodosum**

17b. Disc of lip with yellow patch 18

18a. Petals margins entire; lip rhombic, pubescent; disc not ridged .... **D. aqueum**

18b. Petals margins undulate; lip broadly obovate; disc with two tubercled-papillose ridges...... ....**D. formosum**

Dendrobium anamalayanum Chandrab., V. Chandras. & N.C. Nair, J. Bombay Nat. Hist. Soc. 78: 575. 1981; Seidenfaden in K.M. Matthew, Fl. Tamilnadu Carnatic 3: 1588. 1983.

Type: India: Tamil Nadu, Coimbatore district, Anamalai, Kavarakal, *Chandrabose* 57259 (holo: CAL!).

Etymology: This species was named after the type locality, Anamalai hills, Tamil Nadu, India.

Epiphytic herbs. Pseudobulb greenish-pink, ovoid, covered with membranous sheaths, *ca* 3 × 2 cm. Leaves 2-3, sessile, elliptic-oblong or elliptic-lanceolate, acute, *ca* 7.5 × 2 cm. Inflorescence terminal in leafy condition, *ca* 12 cm long; raceme 5-many-flowered. Floral bracts shorter than the ovary, lanceolate, acuminate, 5-veined, *ca* 9 × 1.5 mm. Pedicel with ovary *ca* 1.1 cm long. Flowers white with pink tinge, *ca* 1.6 cm across. Sepals *ca* 1.1 × 0.4 cm, obtuse, 5-veined; dorsal oblong or oblong-ovate; laterals falcate, oblong-lanceolate; mentum *ca* 2 mm long. Petals oblanceolate, obtuse, 5-veined, *ca* 1.2 × 0.4 mm. Lip broadly obovate, *ca* 9 × 9 mm, 3-lobed, with broad sinuses in between the lobes; laterals ovate-acute; midlobe rhomboid-ovate, margins serrulates; disc 3-toothed. Column white, broadly ovate, 1.5 × 1 mm; foot *ca* 2.5 mm long. Pollinia 4, oblong, 1 mm long. Capsules ellipsoid, *ca* 2 cm long.

Fl. & Fr.: April-September.

Habitat: Rare, it grows on densely moss-covered tree trunks in evergreen forest (above 800 m).

Distribution: Endemic to Tamil Nadu and Kerala. ***Eastern Ghats:*** *TN:* Shevaroy hills.

Specimens Examined: *TN:* Shervarayan temple (SLM), *DIA* 2985, *KMM* 23334 (RHT).

Dendrobium aphyllum (Roxb.) C.E.C. Fisch., Fl. Madras 1416. 1928; Mooney, Suppl. Bot. Bihar & Orissa 213. 1956; Misra, Orchids Orissa 448. 2004; Ramakrishna in Pullaiah, Fl. Andhra Pr. 3: 936. 1997; Misra *et al.,* Bull. Bot. Surv. India 50: 135. 2008. *Limodorum aphyllum* Roxb., Pl. Coromandel 1: 34. 1795. *Dendrobium cucullatum* R. Br., Bot. Reg. 7: t. 548. 1821.

Type: India: Coromandel Coast, Roxburgh icon 241 (CAL & K).

Etymology: The specific epithet refers to the absence of leaves during flowering season.

Epiphytic herbs. Stems pendulous, yellowish-green, sub-clavate, internodes covered with leaf sheaths. Leaves membranous, sessile, lanceolate or elliptic-lanceolate, acuminate, *ca* 8 × 2.5 cm. Inflorescence lateral, on the leaf less nodes, *ca* 2.5 cm long; fascicles 2-3-flowered. Floral bracts brownish, triangular, acute, 5-veined, *ca* 4 × 1.8 mm. Pedicel with ovary greenish-maroon, *ca* 3 cm long. Flowers fragrant, whitish-rose, *ca* 5 cm across. Sepals oblong-lanceolate, acute, 5-veined, *ca* 3.2 × 0.8 cm; mentum conical, *ca* 5 mm long. Petals elliptic-oblong, acute, 5-veined, *ca* 3 × 1 cm. Lip cucullate, sub-panduriform or sub-orbicular, base tubular, margins irregularly dentate, *ca* 3.2 × 3.1 cm, pubescent and with hairs along the margins, 5-veined; disc with 3-strong veins. Column whitish-yellow, oblong, *ca* 4 × 2 mm; foot *ca* 6 mm long. Stigma oblong; rostellum cupulate. Pollinia 4, oblong, *ca* 3 × 2 mm. Capsules oblong-ellipsoid, *ca* 6 cm long.

Fl. & Fr.: March-July.

Habitat: Occasional in semi-evergreen and moist deciduous forests (600-1000 m).

Distribution: ***World*:** Australia, Bangladesh, Bhutan, Borneo, Cambodia, China, India, Java, Laos, Malaysia, Myanmar, Nepal, Sri Lanka, Sulawesi, Sumatra, Thailand and Vietnam. ***India:*** Andaman & Nicobar Is., Andhra Pradesh, Arunachal Pradesh, Assam, Chhattisgarh, Jharkhand, Karnataka, Kerala,

Madhya Pradesh, Maharashtra, Manipur, Meghalaya, Mizoram, Nagaland, Odisha, Sikkim, Tamil Nadu, Tripura, Uttarakhand and West Bengal. ***Eastern Ghats:*** Occasionally found in Eastern Ghats of Andhra Pradesh, Karnataka and Odisha.

Specimens Examined: *AP:* Sunkarimetta (VSKP), *BR & KP* 34650 (SKU); Gurthedu (EG), *SS* 3608 (AU); Gudem valley (VSKP), *VNS* 493 (CAL); G. Madugula (VSKP), *GVS* 47286 (MH); Chintapalli (VSKP) *SM* 1969 (RPRC). *OD:* Similipal forest-Manabhomga ghat (MB), *SM* 90 (CAL); Similipal forest-Baniabasa (MB), *Bahali & Agrawala* 593 (CAL).

Dendrobium aqueum Lindl., Edwards's Bot. Reg. 29(Misc.) 5. 1843; Hook.f., Fl. Brit. India 5: 739. 1888; C.E.C. Fisch., Fl. Madras 1416. 1928; Seidenfaden in K.M. Matthew, Fl. Tamilnadu Carnatic 3: 1589. 1983; Ramakrishna in Pullaiah, Fl. Andhra Pr. 3: 936. 1997; Misra *et al.,* Bull. Bot. Surv. India 50: 135. 2008.

Type: India: Bombay, *Loddiges s.n.* (Probably BLAT?).

Etymology: The specific epithet refers to appearance (large size) of flowers of the species.

Epiphytic herbs. Stems pendulous, yellowish-green when young and brownish at age, sub-clavate, internodes covered with sheathing leaf bases. Leaves membranous, sessile, ovate-lanceolate, acuminate, *ca* 16 × 4 cm. Inflorescence axillary in leafy stem, *ca* 2 cm long; fascicles 2-3-flowered. Floral bracts pale white, ovate-lanceolate, acuminate, *ca* 4 × 2 mm. Pedicel with ovary yellowish-green, *ca* 2.6 cm long. Flowers fragrant, creamish-white with yellow disk. Sepals oblong-ovate, acute or acuminate, 5-7-veined; dorsal *ca* 2.2 × 0.8

cm; laterals *ca* 2 × 1 cm; mentum *ca* 1 cm long. Petals broadly obovate-oblong or elliptic-oblong, acute, 5-veined, *ca* 2.2 × 1.5 cm. Lip rhombic, *ca* 2 cm long, pubescent, 3-lobed; laterals broadly triangular, *ca* 5 mm long; midlobe incurved, broadly triangular, margins erose, acute or obtuse, *ca* 1.2 × 1 cm. Column pale white, oblong, *ca* 4.5 × 2 mm; foot *ca* 1 cm long, rounded. Stigma oblong-orbicular. Pollinia 4, ovoid or pyriform, 1.5 × 0.8 mm. Capsules obovate-pyriform, *ca* 3 cm long.

Fl. & Fr.: September-October.

Habitat: Rare, it grows on densely moss covered tree trunks of moist deciduous and evergreen forests (800-1600 m).

Distribution: Endemic to Peninsular India (Andhra Pradesh, Karnataka, Kerala, Maharashtra and Tamil Nadu). ***Eastern Ghats:*** *AP:* Yarlagadda forest; *TN:* Kolli hills and Shevaroy hills.

Specimens Examined: *AP:* Yarlagadda metta (EG), *VNS* 170 (CAL). *TN:* Kolli hills, Periyashola (NK), *AM* 13238 (RTH); Kolli hills, Kulivalavu (NK), *KMM & CM* 18659, *KMM* 62744 (RHT); Kolli hills, Madal shola (NK), *KMM* 63146, 63186, 63192 & 63364 (RHT); Kolli hills (TRP), *S. Parthipan* and Party *s.n.* (MH); Kollimalai (TRP), *S. India Flora* 11414 (MH); Shevaroy hills (SLM), *AVNR* 18267 (MH); Shevaroy hills (SLM), *DIA* 4168, 7833, 10197 (RHT); Shevaroy hills, Kakasholai (SLM), *KMM & DIA* 4929, *KMM* 23371, 25810 & 25916 (RHT); Shevaroy hills, Yercaud (SLM), *KMM* 63253 (RHT).

Dendrobium bicameratum Lindl., Edwards's Bot. Reg. 25(Misc.): 59. 1839; Hook.f., Fl. Brit. India 5: 729. 1888; Haines, Bot. Bihar & Orissa, 1173. 1924; C.E.C. Fisch., Fl. Madras 1416. 1928; Misra, Orchids Orissa 419. 2004.

Type: India: North India, *Gibsons s.n.* (not located).

Etymology: The specific epithet refers to the bilocular foot of the column.

Epiphytic or lithophytic herbs. Stems narrowly fusiform-clavate. Leaves sessile, elliptic-oblong to elliptic-lanceolate, *ca* 10 × 1 cm. Inflorescence 3-6-flowered, sub-capitate, *ca* 8 mm long. Floral bracts oblong-lanceolate or ovate, *ca* 3.5 mm long. Pedicel with ovary pale green, *ca* 6 mm long. Flowers pale yellow with purple or red speckled, *ca* 8 mm across. Dorsal sepal broadly obovate-oblong or ovate-oblong, rounded, *ca* 7 × 4.4 mm; lateral sepals ovate-triangular, *ca* 6 mm long; mentum urceolate. Petals broadly elliptic, acute, entire or irregularly dentate. Lip fleshy, broadly triangular, *ca* 8 mm long, 3-lobed; laterals triangular, bright purple; midlobe broadly ovate or truncate, margins crispate; disc with cushion-shaped callus. Column broadly ovate, *ca* 1.5 mm long; foot biloculate, *ca* 4 mm long. Pollinia 4, ovoid, 1 mm long. Capsules fusiform-ovoid, *ca* 1.4 cm long.

Fl. & Fr.: August-May.

pseudobulbs greenish-yellow, ridged, fusiform, uni-nodded, *ca* 5 × 1 cm. Leaf sessile, oblong-linear or oblong-lanceolate or narrowly lanceolate, acute, *ca* 10 × 2 cm. Cymes axillary, 1-3-flowered. Floral bracts broadly ovate, acuminate, 3-veined, *ca* 2 × 2.5 mm. Pedicel with ovary yellowish-green, *ca* 6 mm long. Flowers scented, white with red-spotted hypochile lip. Sepals oblong-lanceolate, acute or obtuse, 5-veined, *ca* 10 × 4.5 mm; mentum *ca* 4.5 mm long. Petals lanceolate, acute, 3-veined, *ca* 9.5 × 2.5 mm. Lip reflexed, conduplicate, *ca* 12 × 6 mm, 3-lobed; laterals oblong; midlobe recurved, sub-quadrate or triangular-cordiform, crenulated; disc with two prominent keels, somewhat wavy, which prolong into short sinuate crests. Column oblong-ovate. Pollinia 4, oblong.

Fl.: September-October.

Habitat: Rare, it grows on tree trunks or moss covered rocks of semi-evergreen and moist deciduous forests.

Distribution: Endemic to Peninsular India (Karnataka, Kerala, Maharashtra and Tamil Nadu). ***Eastern Ghats:*** *TN:* Shevaroy hills.

Specimen Examined: *TN:* Shevaroy hills (SLM), *KMM* 25816 (RHT).

Dendrobium peguanum Lindl., J. Proc. Linn. Soc., Bot. 3: 19. 1858; Misra, Orchids Orissa 431. 2004. *D. pygmaeum* Lindl., Gen. Sp. Orchid. Pl. 85. 1830 (illegitimate); Hook.f., Fl. Brit. India 5: 717. 1890; Haines, Bot. Bihar & Orissa, 1173. 1924.

Syntypes: Myanmar: Pegu, *McLelland s.n.* (K (image!)); Borneo: Lobb *s.n.* (K).

Etymology: This species was named after the type locality, Pegu (Myanmar).

Epiphytic herbs. Pseudobulbs in clusters, elliptic-ovate or globular-ovoid. Leaves 2-4, linear-oblong or oblong-lanceolate, *ca* 7 cm long. Racemes lateral or terminal, densely few-flowered. Floral bracts ovate-lanceolate. Pedicel with ovary *ca* 4.5 mm long. Flowers *ca* 1.5 cm across. Sepals white, oblong-lanceolate; mentum funnel-shaped. Petals white, oblanceolate or spathulate. Lip greenish-brown veined with dark brown, anterior lobe purple, shortly clawed, obovate, 3-lobed, midlobe margins strongly crisped; disc with cushion-like, rugose. Pollinia 4, clavate.

Fl.: November-February.

Habitat: Rare, grows on tree trunks of moist deciduous forests (above 400 m).

Distribution: ***World:*** India, Indonesia, Nepal, Myanmar and Thailand. ***India:*** Gujarat, Karnataka, Jharkhand, Maharashtra, Mizoram, Odisha, Sikkim and West Bengal. ***Eastern Ghats:*** *OD:* Govindapali-Gupteswara forest and Similipal forest.

Specimens Examined: *OD:* Matili (KP), *SM* 30 (CAL).

Dendrobium regium Prain, J. Asiat. Soc. Bengal, Pt. 2, Nat. Hist. 71: 80. 1902; Haines, Bot. Bihar & Orissa, 1174. 1924; Mooney, Suppl. Bot. Bihar & Orissa 213. 1956; Misra, Orchids Orissa 423. 2004; Misra *et al.,* Bull. Bot. Surv. India 50: 136. 2008.

Type: India: Jharkhand or Chhattisgarh, *Prain s.n.* (K).

Etymology: The specific epithet refers to beautiful (wonderful) flowers of the species.

Epiphytic or lithophytic herbs, *ca* 45 cm high. Leaves oblong or oblong-lanceolate, *ca* 10 × 1.5 cm. Inflorescence lateral on leafy or leaf less stem, *ca* 10 cm long; raceme laxly 2-4-flowered. Floral bracts lanceolate or oblong, acute, 1 mm. Pedicel with ovary *ca* 6 cm long. Flowers magenta with dark veins and transversely markings, *ca* 8 cm across. Sepals oblong-elliptic, acute; dorsal *ca* 4.5 × 1.5 cm; laterals *ca* 5 × 1.8 cm; mentum conical. Petals ovate, base shortly clawed, acute or obtuse, *ca* 4.6 × 3 cm. Lip shortly clawed, broadly ovate, glabrous, *ca* 4.2 × 3.2 cm. Column oblong-ovate. Pollinia 4, oblong, 2.5 × 1 mm. Capsules narrowly obovoid, *ca* 5 cm long.

Fl. & Fr.: April-October.

Habitat: Rare, grows in semi-evergreen and moist deciduous forests (above 800 m).

Distribution: Endemic to India (Andhra Pradesh, Bihar, Chhattisgarh, Jharkhand and Odisha). ***Eastern Ghats:*** *AP:* Kotapalli forest; *OD:* Rebana forest and Similipal forest.

Specimens Examined: *AP:* Kotapalli (VSKP), *SM* 106 (RPRC). *OD:* Similipal forest-Tinadiha (MB), *SM* 73 & 74 (CAL); Similipal forest-Jenabila (MB), *SM* 1551 (CAL).

Dendrobium transparens Wall. *ex* Lindl., Gen. Sp. Orchid. Pl. 79. 1830; Hook.f., Fl. Brit. India 5: 738. 1890; Haines, Bot. Bihar & Orissa, 1174. 1924; Mooney, Suppl. Bot. Bihar & Orissa 213. 1956; Misra, Orchids Orissa 439. 2004.

Syntypes: Nepal: Wallich *s.n.* in Wall. Cat. no. 2008 (CAL!, K, M, BR, P, BM, E).

Etymology: The specific epithet refers to the thin-textured (transparent) periath of the species.

Epiphytic herbs, *ca* 60 cm high. Stems yellowish, erect, internodes covered with sheathing leaf bases. Leaves membranous, sessile, linear-lanceolate or elliptic-oblong, acute, *ca* 10 × 2 cm. Cymes on leafless stem, 2-3-flowered. Floral bracts greenish-white, broadly lanceolate, acute, *ca* 9 × 6 mm. Pedicel with ovary purplish-white, *ca* 3 cm long. Flowers white suffused with purple and lip with large purple patch at base, *ca* 5 cm across. Sepals lanceolate or oblong-lanceolate, acute, 5-veined; dorsal *ca* 3 × 0.6 cm; laterals *ca* 3.2 × 0.8 cm; mentum conical, *ca* 5 mm long. Petals broadly ovate, acute, 3-veined, *ca* 3 × 1 cm. Lip clawed, convolute at base, broadly elliptic-obovate, *ca* 3.5 × 1.5 cm, obscurely 3-lobed; laterals broad, erose; midlobe sub-orbicular, margins undulate or serrate. Column whitish-pink, oblong, *ca* 3 × 2.5 mm, with arms; foot straight, *ca* 6 mm long. Stigma oblong-elliptical. Pollinia 4, oblong, 1 × 0.8 mm. Capsules ovoid, *ca* 2 cm long.

Fl. & Fr.: March-July.

Habitat: Occasionally it grows on tree trunks of semi-evergreen and moist deciduous forests (above 600 m).

Distribution: ***World:*** Bangladesh, Bhutan, India, Myanmar and Nepal. ***India:*** Arunachal Pradesh, Assam, Chattisgarh, Jharkhand, Manipur, Meghalaya, Mizoram, Nagaland, Odisha, Sikkim, Tripura, Uttarakhand and West Bengal. ***Eastern Ghats:*** *OD:* Gopalpur-Gunupur forest, Govindapali forest, Mahendragiri hills, Rebana forest and Similipal forest.

Specimens Examined: *OD:* Similipal forest-Dhanagoji (MB), *SM* 1544 (CAL); Rebana forest-Meghanada (KJ), *SM* 546 (CAL); Kasipur (KP), *SM* 2 (CAL).

Dendrobium wightii A.D. Hawkes & A.H. Heller, Orquídea (Rio de Janeiro) 24: 16. 1962. *D. graminifolium* Wight, Ic. Pl. Ind. Or. t. 1649, 1851 (auct. non Willd.); Hook.f. Fl. Brit. Ind. 5: 718. 1890; C.E.C. Fisch., Fl. Madras 1415. 1928. *D. wightii* M.S. Balakr., J. Bombay Nat. Hist. Soc., 67: 66. 1970 (Illegitimate); Seidenfaden in K.M. Matthew, Fl. Tamilnadu Carnatic 3: 1590. 1983.

Syntypes: India: Tamil Nadu, Cutrallum, *Wight.* 904 & 2954 (CAL! & K (images!)).

Etymology: This species was named in honour of R. Wight (1796-1872), well known plant collector from Peninsular India and also collected type material.

Epiphytic or lithophytic herbs, *ca* 20 cm high. Rhizome creeping, jointed. Stems erect, elongated, internodes covered with sheathing leaf bases. Leaves grass like, narrowly linear or lanceolate, acute, *ca* 8 × 0.6 cm. Inflorescence terminal, filiform, flexuose, *ca* 5 cm long; raceme laxly 2-6-flowered. Floral bracts greenish-purple, ovate, acute, veined, *ca* 2.5 × 1.5 mm. Pedicel with ovary pale pink, *ca* 1 cm long. Flowers fragrant. Sepals whitish-pink, oblong-lanceolate, acute, 3-5-veined; dorsal *ca* 8 × 2 mm; laterals *ca* 8.5 × 3 mm; mentum conical, *ca* 5 mm long. Petals whitish-pink, narrowly lanceolate or oblanceolate, acute, 3-veined, *ca* 7.5 × 1.2 mm. Lip white with green tinge at base and pink tinge at near the apex or not, clawed, obovate, *ca* 1.2 cm long, 3-lobed; laterals blunt, triangular; midlobe flat, orbicular-ovate, crenulated, *ca* 6 × 4.5 mm; disc flat. Column white, broadly oblong, *ca* 2 × 2.4 mm, with apical arms; foot oblong, *ca* 4 mm long. Stigma oblong. Pollinia 4, oblong, 1.1 × 0.8 mm. Capsules ovoid, *ca* 8 mm long.

Fl. & Fr.: July-December.

Habitat: Rare, it grows on fully moss covered tree trunks or rock of evergreen forests (above 1000 m).

Distribution: Endemic to South India (Karnataka, Kerala and Tamil Nadu). ***Eastern Ghats:*** *TN:* Shevaroy hills.

Specimen Examined: *TN:* Shevaroy hills, Sanyasi malai (SLM), *AVNR* 23067 (MH).

Doubtful species

Dendrobium fimbriatum Hook., Exot. Fl. 1: t. 71. 1823; Hook. f., Fl. Brit. India 5: 745. 1888; Haines, Bot. Bihar & Orissa, 1175. 1924; Mooney, Suppl. Bot. Bihar & Orissa 213. 1956; Misra, Orchids Orissa 431. 2004.

Note: This species was reported by Mooney from Bissam Cuttak, Near Water fall and no herbarium material support this report and its occurrences in Odisha expressed doubtful by Misra 2004. Based above record, Reddy *et al.* (2005) reported from the Eastern Ghats of Odisha. Futher explorations are needed for its occurrence confirm in Eastern Ghats.

Dendrobium ovatum (L.) Kraenzl., Pflanzenr. IV, 50 II B 21: 71. 1910; C.E.C. Fisch., Fl. Madras 1416. 1928; Ramakrishna in Pullaiah, Fl. Andhra Pr. 3: 937. 1997; Misra *et al.,* Bull. Bot. Surv. India 50: 136. 2008. *Epidendrum ovatum* L., Sp. Pl. 952. 1753. *Dendrobium chlorops* Lindl., Edwards's Bot. Reg. 30 (Misc.): 44. 1844; Hook.f., Fl. Brit. India 5: 719. 1888.

Note: This species was reported from Papavinasanam valley (Tirumala) of Seshachalam hills. Based on above report, Ramakrishna (1997), Reddy *et al.* (2005), Misra *et al.* (2008) recorded its occurrence in the Eastern Ghats of Seshachalam hills. We collected this species in the same locality in flowering condition and identified it as a *Dendrobium macrostachyum* and we presume that it was misidentified earlier as *Dendrobium ovatum*. Further, *Dendrobium ovatum* is endemic to Western Ghats.

Didymoplexis Griff.

Calcutta J. Nat. Hist. 4: 383. t. 17. 1844.

World:20/India:3/EG:1

Type: *Didymoplexis pallens* Griff.

Etymology: The generic name is derived from Greek words, *didymo* (double) and *plexis* (unite), refers to the fused nature of sepals and petals (tube).

Distribution: Tropical and South Africa to W. Pacific.

Didymoplexis pallens Griff., Calcutta J. Nat. Hist. 4: 383. t. 17. 1844; Hook.f., Fl. Brit. India 6: 122. 1890; C.E.C. Fisch., Fl. Madras 1459. 1928; Misra, Orchids Orissa 267. 2004; Ramakrishna in Pullaiah, Fl. Andhra Pr. 3: 937. 1997; Misra *et al.,* Bull. Bot. Surv. India 50: 136. 2008.

Type: India: West Bengal, Sri Rampur (Serampur), *Griffith s.n.* (holo: CAL).

Etymology: The specific epithet refers to the colour of the flowers of the species.

Saprophytic, leafless herbs, *ca* 20 cm high. Roots fasciculate; tubers ellipsoid or fusifrom, often moniliform, nodded. Stem subterranean, ridged, covered with tubular sheaths. Inflorescence *ca* 15 cm long; raceme laxly 2-6-flowered. Floral bracts brown, broadly triangular, 3-veined, *ca* 4 mm long. Pedicel with ovary brownish-green, *ca* 1 cm long. Flowers delicate, white with yellow tinge on the lip. Sepals and petals fused at base and form sub-campanulate 5-lobed tube; tube *ca* 3 mm long; lobes of dorsal sepal and petals oblong-obovate or ovate-deltoid, 3-veined; lateral lobes reflexed, suborbicular-oblong. Lip include in tube, stipitate, deltoid, truncate, erose-crenate, *ca* 6 mm long; disk with stipitate-warty papillose, with irregular yellow calli. Column white, clavate, winged; foot short, incurved. Stigma suborbicular-squarish. Pollinia 4, obliquely oblong, 0.8 mm. Capsules ellipsoid or fusiform, *ca* 3 cm long.

Fl. & Fr.: June-July.

Habitat: Rare, it grows under dense shady conditions in humus rich soils of moist deciduous forest (above 500 m).

Distribution: ***World*:** Afghanistan, Australia, Bangladesh, Bhutan, Borneo, India, Japan, Java, Malaysia, Myanmar, New Guinea, Pacific Is., Philippines, Sri Lanka, Sumatra, Timor, Thailand and Vanuatu. ***India:*** Andhra Pradesh, Arunachal Pradesh, Assam, Karnataka, Kerala, Maharashtra, Meghalaya, Odisha, Sikkim, Tamil Nadu and West Bengal. ***Eastern Ghats:*** *AP:* Godavari; *OD:* Gopinathpur forest and Mahendragiri hills.

Specimens Exmined: *OD:* Siddheswari hills (GJM), *SM* 2371, 2372, 2373, 2374 (CAL); Gopinathpur (GJM), *SM* 2384, 2387, 2388 (CAL).

Dienia Lindl.

Bot. Reg. 10: t. 825. 1824.

World:2/India:1/EG:1

Type: *Dienia congesta* Lindl.

Etymology: The generic name is derived from greek words, *di* (two) and *annusus* (a year), refers to two years old (life span of each growth of the plant).

Distribution: Tropical & Subtropical Asia to West Pacific.

Dienia ophrydis (J. Konig) Seidenf., Contr. Orchid Fl. Thailand 13: 18. 1997; Misra, Orchids Orissa 375. 2004. *Epidendrum ophrydis* J. Konig, Observ. Bot. 6: 46. 1791. *Malaxis ophrydis* (J. Koenig) Ormerod in G. Seidenfaden, Descr. Epidendrorum J.G. Konig 18. 1995. *Microstylis congesta* (Lindl.) Rchb.f., Ann. Bot. Syst. 6: 206. 1861; Hook.f., Fl. Brit. India 5: 689. 1888; Haines, Bot. Bihar & Orissa, 1165. 1924.

Types: Thailand: *Konig s.n.* (LIV & K).

Etymology: This species shows resemblance to the orchid genus *Ophrys*.

Terrestrial robust herbs, *ca* 30 cm high. Rhizome creeping, covered with tubular scales; pseudobulbs green, fusiform-oblong, nodded, covered with sheaths, *ca* 8 × 1.5 cm. Leaves 3-6, membranous, plicate, ovate to oblong-ovate or ovate-lanceolate, margins undulate, apex acuminate, *ca* 20 × 6 cm. Inflorescence *ca* 30 cm long, slender, rachis ribbed; raceme cylindrical, densely many-flowered. Floral bracts longer than the ovary, deflexed, narrowly lanceolate, acuminate, *ca* 6 × 1 mm. Pedicel with ovary green, *ca* 4 mm long. Flowers non-resupinate, maroon, *ca* 5 mm across, gland dotted. Sepals deflexed, oblong or oblong-ovate, acute, margin revolute, 1-3-veined, *ca* 3 × 1.1 mm. Petals inflexed, linear-oblanceolate or linear-spathulate, acute, 1-veined, *ca* 3 × 0.8 mm. Lip fleshy, sub-quadrately-ovate, *ca* 3 × 1.5 mm, 3-lobed at apex; laterals blunt, broad, 0.8 mm long; midlobe oblong-ovate, obtuse, 1 × 0.5 mm; callus at base, erect. Column maroon, oblong, 1.5 × 1 mm. Stigma transversely elliptic. Pollinia 4, oblong-ovoid, 0.4 × 0.4 mm. Capsules oblong-clavate, *ca* 6 mm long.

Fl. & Fr.: June-August.

Habitat: Occasional, it grows under dense shady condition in humus soils of semi evergreen and moist deciduous forests (500-1000 m).

Distribution: ***World*:** Australia, Bangladesh, Bhutan, Borneo, Cambodia, China, India, Java, Laos, lesser Sunda Is., Malaysia, Moluccas, Myanmar, Nepal, New Caledonia, New Guinea, Philippines Sri Lanka, Sulawesi, Sumatra, Taiwan, Thailand and Vietnam. ***India:*** Andaman and Nicobar Is., Andhra Pradesh, Arunachal Pradesh, Assam, Bihar, Chhattisgarh, Jharkhand, Kerala, Manipur, Meghalaya, Mizoram, Nagaland, Odisha, Sikkim, Tamil Nadu, Tripura and West Bengal. ***Eastern Ghats:*** *AP:* Cherukonda; *OD:* Rebana forest and Similipal forest.

Specimens Exmined: *AP*: Gudem (VSKP), *BR* & *KP* 41621 (SKU); Sambarukonda (VSKP), *BR, KP* & *MVS* 34691 (SKU); Cherukonda (VSKP),

CSR 1892 (KU). *OD*: Similipal forest-Meghasani (MB), *SM* 1678 (CAL); Similipal-Udiabasa (MB), *SM* 1526 (CAL); Rebana-Panasapani (KJ), *SM* 733 (CAL); Upper Barakamara –Jenabil, Similipal RF (MBJ), *sine coll.* 1408 (BSID).

Diplocentrum Lindl.

Edwards's Bot. Reg. 18: t. 1522. 1832.

World:2/India:2/EG:1

Type: *Diplocentrum recurvum* Lindl.

Etymology: The generic name is derived from Greek words, *diploos* (double) and *kentron* (spur), refers to the two spurrs at the base of the lip.

Distribution: South India and Sri Lanka.

Diplocentrum recurvum Lindl., Edwards's Bot. Reg. 18: t. 1522. 1832; Hook.f., Fl. Brit. India 6: 78. 1890; C.E.C. Fisch., Fl. Madras 1449. 1928; Seidenfaden in K.M. Matthew, Fl. Tamilnadu Carnatic 3: 1602. 1983; Ramakrishna in Pullaiah, Fl. Andhra Pr. 3: 938. 1997; Misra *et al.*, Bull. Bot. Surv. India 50: 136. 2008.

Type: India: Indian Orientalis, Herb. *Heyne* in Wall. Cat. no. 7331 (CAL!).

Etymology: The specific epithet refers to the recurved nature of the leaves of the species.

Epiphytic herbs. Stems erect or pendulous, stout, terete, *ca* 20 cm long, covered with sheathing leaf bases. Leaves coriaceous, recurved, sessile, linear or linear-oblong, unequally bilobed, *ca* 12 × 1.5 cm, green flushed with pink. Inflorescence axillary, erect, longer than leaves, *ca* 25 cm long; panicles laxly many-flowered. Floral bracts brown, triangular, acute, 1 × 1 mm. Pedicel with ovary greenish-pink, *ca* 6 mm long. Flowers fragrant, rose-pink or brownish tinged with pink, *ca* 6 mm across. Dorsal sepal hooded, oblong-ovate, acute, 1-3-veined, *ca* 3 × 1.8 mm; lateral sepals falcate, obovate-oblong or lanceolate, acute, *ca* 3.5 × 2.2 mm. Petals oblong-ovate, acute, 1-veined, *ca* 2.6 × 1.8 mm. Lip sessile, oblong-ovate or lingulate, margins curved, 2-spurred, *ca* 4.5 × 2 mm; disc with median thick fleshy ridge; spurs collateral, conical, incurved, *ca* 2 mm long. Column whitish-rose, clavate, 1 × 1 mm, without foot. Stigma squarish-orbicular. Pollinia 2, ovoid, 0.6 × 0.5 mm. Capsules pyriform, *ca* 1.6 cm long, with marcescent perianth.

Fl. & Fr.: May-July.

Habitat: Rarely it grows on fully moss covered tree trunks of evergreen forest (above 800 m).

Distribution: ***World***: India and Sri Lanka. ***India:*** Andhra Pradesh, Karnataka, Kerala and Tamil Nadu. ***Eastern Ghats:*** *TN:* Shevaroy hills and Kolli hills.

Specimens Examined: *TN:* Shevaroy hills, near Orchidarium (SLM), *AVNR* 23061 (MH); Shevaroy hills (SLM), *E. Barnes s.n.* (DD); Shervaroyan temple (SLM), *NV* & *TSJ* 22455, *KMM* 23342 (RHT); Shevaroy hills, Yercaud (SLM), *KMM* 30452 & 30465 (RHT); Kollimalai (TRP), *CAB* 11367, 13023 (MH); Kolli hills (NK), *KMM* 62264, 62656, 62726 & 63772 (RHT), *AM* 11663, 13176 & 13274 (RHT); Sirumalai (DGL), *J. Pallithanam* 53394 & 53395 (RHT); Bodamalais (SLM), *KMM* & *KM* 14873, *KMM* 15435, *CP* & *CM* 2215 (RHT); Guthirayan hills (DMP), *KMM* & *NV* 23964 (RHT); Yelagiri hills (VLR), *KMM* & *KTA* 61055 (RHT).

Diploprora Hook.f.

Fl. Brit. India 6: 26. 1890.

World: 2/India:2/EG:1

Type: *Diploprora championii* (Lindl.) Hook.f.

Etymology: The generic name is derived from the Greek words, *diplo* (two) and *prora* (prow), refers to the bifurcation at apex of lip.

Distribution: India, Sri Lanka, Himalaya to Taiwan.

Diploprora championii (Lindl.) Hook.f., Fl. Brit. India 6: 26. 1890; Misra, Orchids Orissa 568. 2004. *Cottonia championii* Lindl., Hooker's J. Bot. Kew Gard. Misc. 7: 35. 1855.

Types: Hong Kong: Victoria Peak, *Champion* 277 (K (images!)).

Etymology: The species was named in honour of John George Champion (1815-1854), who collected type specimens.

Epiphytic herbs, *ca* 12 cm high. Stem pendant, ascending at the distal half, terete, covered with sheathing leaf bases, nodes with roots. Leaves coriaceous, sessile, falcate, oblong-lanceolate to ovate-elliptic, acute, *ca* 10 × 2 cm. Inflorescence leaf-opposed, flexuous, *ca* 6 cm long; raceme laxly few-flowered. Floral bracts greenish-yellow, ovate, acute, 7-veined, *ca* 4.5 × 2.8 mm. Pedicel with ovary yellowish, *ca* 6 mm long. Flowers *ca* 1.3 cm across. Sepals and petals yellow, keeled, obtuse, 5-7-veined; dorsal sepal oblong-elliptic or ovate-elliptic, *ca* 9 × 5 mm; lateral sepals ovate, *ca* 10 × 5 mm; petals oblong-oblanceolate or lanceolate-elliptic or obovate, *ca* 9 × 4 mm. Lip pale white with yellow or reddish-purple markings within, sigmoidally curved, *ca* 8 × 4 mm; hypochile boat-shaped; epichile incurved, base broadly triangular, with 2 long caudate lobes at apex; disc with keel. Column white or yellow, oblong, *ca* 3 × 2.8 mm. Stigma trapezoid; rostellum deeply notched. Pollinia 2, globose, 0.8 × 0.8 mm. Capsules cylindric-fusiform, *ca* 5 cm long.

Fl. & Fr.: June-October.

Habitat: Rare in evergreen stunted forest (above 1000 m).

Distribution: ***World:*** Bhutan, China, India, Myanmar, Sri Lanka, Taiwan, Thailand and Vietnam. ***India:*** Andaman & Nicobar Is., Arunachal Pradesh, Assam, Karnataka, Meghalaya, Sikkim and Odisha. ***Eastern Ghats:*** *OD:* Similipal forest.

Specimens Exmined: *OD:* Similipal forest-Garandia (MB), *SM* 1287 (CAL).

Disperis Sw.

Kongl. Vetensk. Acad. Nya Handl. 21: 218. 1800.

World:79/India:3/EG:1

Type: *Disperis secunda* Sw. (lectotype).

Etymology: The generic name is derived from Greek words, *disperis* (two-pouched) and *dis-phra* (the two anther loculi), refers to the nature of anther.

Distribution: Africa, Tropical and Subtropical Asia to NW. Pacific

Disperis zeylanica Trimen, J. Bot. 23: 245. 1885. *D. neilgherrensis* auct non Wight, 1851; C.E.C. Fisch., Fl. Madras 1478. 1928; Seidenfaden in K.M. Matthew, Fl. Tamilnadu Carnatic 3: 1569. 1983.

Type: Sri Lanka: Central Province, *Thwaites s.n.* (K).

Etymology: This species was named after type locality country, Sri Lankan old name, Zeylan.

Terrestrial herbs, *ca* 15 cm high. Tubers 1-2, ovoid, puberulent. Stem dark green to purple, terete. Leaves 1-3, cordate-amplexicaul, acute, *ca* 1.5 × 2 cm. Rachis 1-3-flowered, pinkish-green. Floral bracts green, foliaceous, ovate, acute, *ca* 1 × 0.5 cm. Ovary green, *ca* 1 cm long. Flowers pink with purple striations. Dorsal sepal hooded, lanceolate, acute, 3-veined, *ca* 6 × 1.5 mm; lateral sepals deflexed, connate at base, ovate or obovate, acute, *ca* 6 × 4.5 mm. Petals obovate, acute, 3-veined, *ca* 6 × 4.5 mm. Lip whitish-pink, clawed, Y-shaped, *ca* 5 mm long, 3-lobed; claw adnate to base of column; lateral lobes spreading, linear; midlobe linear, terminating into swollen papillose appendage. Stigma lobes separated and located on each side of rostellum; rostellum erect, rhomboid, with 2 curved or coiled arms. Anther dorsal, attached to column by short filament; pollinia 2, curved, oblong. Capsules cylindrical, *ca* 1.5 cm long.

Fl.: June-August.

Habitat: Rare, it grows in high altitude open grass lands and evergreen forest (above 1200 m).

Distribution: ***World:*** India and Sri Lanka. ***India*:** Karnataka, Kerala and Tamil Nadu. ***Eastern Ghats:*** TN: Kolli hills and Shevaroy hills.

Specimens Examined: *TN:* Shevaroy hills, Yercaud (SLM), *Perrottet* 469 (CAL); Shevaroy hills-Nagalur road-Yercaud (SLM), *EV* 77724 (MH), *S. Karthikeyan* 28290 (MH); Shevaroy hills, Yercaud (SLM), *KMM* 23412 (RHT); Shevaroy hills, Manjakottai (SLM), *KMM* 23442 (RHT); Kolli hills (NK), *KMM & CP* 24131, *KMM* 30429, 30436 & 30440 (RHT); Kolli hills (NK), *KRK & RVS* 52165 (FRLH) & *MBV* 17167 (FRLH); Sothiparai (NK), *KMM* 30414 & 62601 (RHT).

Epipogium J.F. Gmel. *ex* Borkh.

Tent. Disp. Pl. German. 139. 1792.

World:4/India:2/EG:1

Type: *Epipogium aphyllum* Sw.

Etymology: The generic name is derived from the Greek words, *epi* (upon) and *pogon* (a beard), refers to the bearded or papillose nature of the disc.

Distribution: Temperate Eurasia, Tropical Africa to SW. Pacific.

Epipogium roseum (D. Don) Lindl., J. Proc. Linn. Soc., Bot. 1: 177. 1857. *Limodorum roseum* D. Don, Prodr. Fl. Nepal. 30. 1825; Murugan *et al.,* Indian J. Forestry 18(2):174. 1995. *Epigonium nutans* (Blume) Rchb.f., Bonplandia 5: 36. 1857; Hook.f., Fl. Brit. India 6: 124. 1890; C.E.C. Fisch., Fl. Madras 1460. 1928.

Type: Nepal: *Wallich s.n.* (holo : BM).

Etymology: The specific epithet refers to the rose coloured flowers of the species.

Saprophytic leaf less herbs, *ca* 60 cm high. Tubers oblong, *ca* 7 × 3.5 cm. Inflorescence *ca* 30 cm long; peduncle with many sheathing sterile bracts; raceme laxly many-flowered. Floral bracts shorter than the ovary, pale white with

(NK), *DIA* 7679, *AM* 11669, 11670 & 13277, *KMM* 62281, 62658 & 63625 (RHT).

Eulophia R. Br.

Bot. Reg. 7: t. 573. 1821 (*nom. cons.*).

World: 201/India:24/EG:7

Type: *Eulophia guineensis* Lindl. (*typ. cons.*).

Etymology: The generic name is derived from Greek words, *eu* (well-developed) and *lophos* (crest), refers to the prominent calluses on the lip.

Distribution: Tropical and Subtropical.

Key to the species

1a. Pseudobulbs epigeal (above the ground), green or greenish-yellow 2

1b. Pseudobulbs hypogeal (below the ground)............ 4

2a. Pseudobulbs fusiform; leaves broad; raceme lateral............**E. pulchra**

2b. Pseudobulbs not as above; leaves grass like; racemes arising from nodes of pseudobulbs 3

3a. Flowers *ca* 2.5 cm across; sepals and petals more than 1.5 cm long.....**E. epidendraea**

3b. Flowers *ca* 1.2 cm across; sepals and petals less than 1.4 cm long............**E. graminea**

4a. Leaves appear after the flowering; inflorescence more the 60 cm long .... **E. flava**

4b. Leaves appear during the flowering; inflorescence less than 60 cm long 5

5a. Flowers perfectly yellow; lip simple ....**E. ochreata**

5b. Flowers not as above; lip 3-lobed............ 6

6a. Flowers greenish-yellow with longitudinal maroon lines and dots; lip sub-panduriform; column without foot ....**E. explanata**

6b. Flowers greenish or greenish-maroon; lip oblong; column with foot ....**E. spectabilis**

Eulophia epidendraea (J. Konig *ex* Retz.) C.E.C. Fisch., Fl. Madras 1434. 1928; Seidenfaden in K.M. Matthew, Fl. Tamilnadu Carnatic 3: 1605. 1983; Ramakrishna in Pullaiah, Fl. Andhra Pr. 3: 939. 1997; Misra *et al.,* Bull. Bot. Surv. India 50: 136. 2008. *Serapias epidendraea* J. Koenig ex Retz., Observ. Bot. 6: 65. 1791. *Eulophia virens* (Roxb.) R. Br. *ex* Lindl., Edw. Bot. Reg. 7: sub. t. 573. 1821; Hook.f., Fl. Brit. India 6: 1. 1890. *Limodorum virens* Roxb., Pl. Coromandel 1: 32. 1795.

Types: India: Tamil Nadu, Trangambar, *Koenig s.n.* (BM & C (images!)).

Etymology: The specific epithet refers to species resemblance with the genus *Epidendrum* L.

Terrestrial herbs. Pseudobulbs epigeal, conico-obpyriform, vertically furrowed, nodded, *ca* 10 × 7 cm. Leaves grass like, plicate, sessile, linear, acuminate, *ca* 80 × 2.5 cm, green and covered with grey coat when fresh. Inflorescence arising from the nodes on the pseudobulb, covered with greyish coat, simple or branched, *ca* 1 m long; raceme laxly many-flowered. Floral bracts shorter than the ovary, pale green, oblong-lanceolate, acuminate, 3-veined, *ca* 1 × 0.3 cm. Pedicel with ovary light green, *ca* 1.5 cm long. Flowers *ca* 2.5 cm across. Sepals and petals pale green with maroon or red veins, reflexed on distal halves, acute or acuminate, 3-veined, veins branched; sepals oblong-lanceolate or linear-lanceolate, *ca* 2.1 × 0.4 cm; petals broadly elliptic, *ca* 1.8 × 0.5 cm. Lip white with purple or red veins, sessile, obovate-oblong, *ca* 2 × 1.2 cm, 3-veined, lateral veins branched, 3-lobed; laterals triangular; midlobe recurved, broadly obovate, rounded, crenulated; disc with 5 ridges, distal central patch sparsely covered with pinkish-white, fat, curved hairs; spur laterally compressed, cylindrical, bulbous at apex, *ca* 6 mm long. Column greenish-yellow, obovate-oblong, *ca* 1 × 0.4 cm, without foot. Stigma sub-quadrately-oblong. Pollinia 2, sub-globose, 2 × 2 mm. Capsules fusiform, *ca* 4 cm long.

Fl. & Fr.: February-September.

Habitat: Rare, it grows in scrub and dry deciduous forests (above 300 m).

Distribution: ***World*:** Bangladesh, India and Sri Lanka. ***India:*** Andhra Pradesh, Karnataka, Kerala, Maharashtra, Tamil Nadu and West Bengal. ***Eastern***

Gahts: *AP:* Seshachalam hills, Veligonda hills and Nattavaram; *TN:* Rare in Eastern Ghats of Tamil Nadu.

Specimens Examined: *AP:* Kalasamudram (ATP), *KRKS* 41865 (SKU); Balapalli (KDP), *JLE* 15744 (MH); Talakona (CTR), *GVS* 27798 (MH); Chittoor, *PS* 3408 (PCM); Veligonda hills (NLR), *MSR* 1390 (CAL); Venkatagiri (NLR), *ASR* 8738; Venkatagiri (NLR), *BSN* 3887 (VV); Nattavaram (VSKP), *CAB* 1903 (MH). *TN:* Gingee RF (SA), *KMS* 12378 (MH); Vandalur RF (CPT), *ANH* 47134 (MH); Karnatigur R.F. (TVM), *RVS* 59174 (FRLH); Kolli hills, Solakkadu (NK), *KMM* & *VKN* 9663 (RHT); Devarbetta (DMP), *KMM* 62935 (RHT).

Eulophia explanata Lindl., Gen. Sp. Orchid. Pl. 180. 1833; Hook.f., Fl. Brit. India 6: 3. 1890; Haines, Bot. Bihar & Orissa, 1171. 1924; Mooney, Suppl. Bot. Bihar & Orissa 211. 1956; Misra, Orchids Orissa 534. 2004; Ramakrishna in Pullaiah, Fl. Andhra Pr. 3: 939. 1997; Misra *et al.,* Bull. Bot. Surv. India 50: 136. 2008.

Type: Sri Lanka: *Macrae s.n.* (holo: K (image!)).

Etymology: The specific epithet refers to spreading nature of the perianth.

Terrestrial herbs. Corms annular, in chain, brownish-white, ovoid, *ca* 3 × 2 cm; stem stout, *ca* 5 cm long. Leaves 2-3, subplicate, broadly elliptic or ovate-elliptic, acuminate, *ca* 20 × 5 cm; petiole *ca* 4 cm long. Inflorescence lateral from the base of corms, simple or branched, *ca* 35 cm long; raceme laxly few-flowered. Floral bracts shorter than the ovary, greenish-maroon, lanceolate, acuminate, 5-veined, *ca* 7 × 1.8 mm. Pedicel with ovary greenish-maroon, *ca* 1.5 cm long. Flowers greenish-yellow with longitudinal maroon lines and dots, *ca* 2 cm across. Sepals spreading, acuminate, 5-veined; dorsal oblanceolate, *ca* 10 × 2.5 mm; laterals oblong-ovate, *ca* 10 × 2.8 mm. Petals spreading, broadly elliptic-ovate, obtuse, 5-veined, *ca* 10 × 5 mm. Lip sessile, sub-panduriform, margin sparingly denticulate, *ca* 12 × 6 mm, 3-lobed; laterals broadly triangular or quadrate; midlobe ovate-oblong, truncately emarginated; disc with 2 crenate lamellae and ridged on distal half; sac laterally compressed, conical, *ca* 4 mm long. Column pale yellow, oblong-ovate, *ca* 4 × 2.2 mm, without foot. Stigma sub-quadrate. Pollinia 2, sub-globose, 1.4 × 1.2 mm. Capsules spindle-shaped, *ca* 4 cm long.

Fl. & Fr.: June-October.

Habitat: Rare, it grows under partially shady or open conditions in humus soils of moist deciduous forest (above 500 m).

Distribution: ***World*:** India and Nepal. ***India:*** Andhra Pradesh, Bihar, Chhattisgarh, Jharkhand, Kerala, Madhya Pradesh, Meghalaya, Odisha, Uttar Pradesh, Uttarakhand and West Bengal. ***Eastern Ghats:*** *AP:* Cheedipalem forest; *OD:* Gopalpur-Gunupur forest and Similipal forest.

Specimens Examined: *AP:* Nichelakanuma (KDP), *BSS* & *SKB* 31067 (SKU); Cheedipalem (EG), *GVS* 27439 (MH). *OD:* Similipal forest-Bamanghaty (MB), *SM* 1642, 1859, 1860 (CAL).

Eulophia flava (Lindl.) Hook.f., Fl. Brit. India 6: 7. 1890. *Cyrtopera flava* Lindl., Gen. Sp. Orchid. Pl. 189. 1833. *C. cullenii* Wight, Icon. Pl. Ind. Orient. 5: t. 1754. 1851. *Eulophia cullenii* (Wight) Blume, Coll. Orchid. 182. 1859. *E. cullenii* (Wight) C.E.C. Fisch., Fl. Madras 1435. 1928 (*nom. nud.*).

Types: Nepal: Mt. Morang, Wall. Cat. no. 7364 (K (images!)).

Etymology: The specific epithet refers to the yellow flowers of the species.

Terrestrial herbs, *ca* 1 m high. Corms in chain, brownish-white, flattened, oval or elliptic, *ca* 8 × 5 cm. Leaves appear after the flowering, 2-3, plicate, sessile, elliptic-lanceolate, acuminate, *ca* 80 × 10 cm. Inflorescence lateral from the base of the corms, simple, *ca* 80 cm long; raceme laxly many-flowered. Floral bracts shorter than the ovary, greenish-yellow, linear-lanceolate, acuminate, 3-veined, *ca* 1.5 × 0.4 cm. Pedicel with ovary greenish-yellow, *ca* 2.6 cm long. Flowers sweet scented, bright yellow, *ca* 4 cm across. Sepals acute, 5-8-

veined, lateral veins branched; dorsal sepal oblong-lanceolate, *ca* 2.6 × 1 cm; laterals elliptic-oblong, *ca* 2.8 × 1 cm. Petals broadly obovate or rhomboid, obtuse or rounded, 3-veined, lateral veins branched, *ca* 2.2 × 1.4 cm. Lip sessile, broadly ovate, *ca* 2.5 × 2.5 cm, 3-lobed above the middle; laterals broadly triangular or orbicular, rounded; midlobe spreading, obovate, crenulated, rounded or emarginated; disc with three prominent nerve fringes and ornamented with maroon dots on striations; spur green, conical, *ca* 7 mm long, obtuse. Column yellowish-white, clavate, *ca* 1.8 × 0.4 cm; foot *ca* 4 mm long. Stigma rectangular-obovate. Pollinia 2, sub-globose, 4 × 3.5 mm. Capsules fusiform, *ca* 4 cm long.

Fl. & Fr.: June-August.

Habitat: Rare, it grows under shady conditions in humus-rich soils of moist deciduous forests (600-800 m).

Distribution: ***World*:** Cambodia, China, Hainan, India, Laos, Myanmar, Nepal, Thailand and Vietnam. ***India:*** Andhra Pradesh, Assam, Kerala, Madhya Pradesh, Maharashtra, Tamil Nadu, Uttar Pradesh and Uttarakhand. ***Eastern Gahts:*** *AP:* Seshachalam hills.

Specimens Examined: *AP:* Matta (KNL), *BSS* & *SKB* 32353 (SKU); Seshachalam hills, Nichenala bodu (KDP), *BR* & *BSS* 30564 (SKU); Nichela kanuma (KDP), *BSS* & *SKB* 31073 (SKU); Pincha hills (KDP), *BSS* & *SKB* 30564, 30565 (SKU); Dongabanda (KDP), *BSS* & *SKB* 30565 (SKU); Talkona (CTR), *BR* & *BSS* 31068 (SKU).

Eulophia graminea Lindl., Gen. Sp. Orchid. Pl. 182. 1833; Hook.f., Fl. Brit. India 6: 2. 1890; C.E.C. Fisch., Fl. Madras 1435. 1928; Misra, Orchids Orissa 542. 2004; Seidenfaden in K.M. Matthew, Fl. Tamilnadu Carnatic 3: 1606. 1983; Ramakrishna in Pullaiah, Fl. Andhra Pr. 3: 939. 1997; Misra *et al.,* Bull. Bot. Surv. India 50: 136. 2008.

Types: Singapore (Singapur): Wall. Cat. no. 7372 (CAL! & K (images!)).

Etymology: The specific epithet refers to the grass-like leaves of the species.

Terrestrial herbs. Pseudobulbs epigeal, bright green, ovoid-conical or pear-shaped, vertically furrowed feebly, *ca* 9 × 2 cm; stem stout, *ca* 4 cm long. Leaves grass like, plicate, sessile, linear, acuminate, green covered with greyish coat when fresh, *ca* 40 × 2 cm. Inflorescence lateral arising from the nodes of pseudobulbs, simple or branched, *ca* 80 cm long; raceme laxly many-flowered. Floral bracts shorter than the ovary, pale green, lanceolate, acuminate, 3-5-veined, *ca* 6 × 1 mm. Pedicel with ovary greenish-pink, *ca* 1.5 cm long. Flowers *ca* 1.2 cm across. Sepals and petals pale green with maroon veins, reflexed, acute or acuminate, 3-5-veined; dorsal oblong-lanceolate, *ca* 1.3 × 0.2 cm; laterals falcate, oblanceolate, *ca* 1.4 × 0.3 cm; petals elliptic-lanceolate, *ca* 1.1 × 0.3 cm. Lip white with purple veins, sessile, fleshy, obovate, *ca* 1.2 × 0.6 cm, 3-veined, lateral veins branched, 3-lobed; laterals triangular; midlobe recurved, obovate, emarginated, crenulate; disc with ridges, central one sparsely covered with pinkish-white, flat, curved hairs; spur maroon-green, dorsiventrally compressed, conical, bulbous at apex, *ca* 3 mm long. Column light green, obovate-oblong, *ca* 4 × 1 mm, without foot. Stigma sub-quadrate. Pollinia 2, sub-globose, 1 × 1.1 mm. Capsules clavate-fusiform, *ca* 4 cm long.

Fl. & Fr.: January-June.

Habitat: Occasional, it grows in open condition of moist, dry deciduous and scrub forests (above 100 m).

Distribution: ***World***: Bangladesh, Bhutan, Borneo, China, India, Japan, Java, Laos, Malaysia, Myanmar, Nepal, Pakistan, Philippines, Ryukyu Is., Singapore, Sri Lanka, Sumatra, Taiwan, Thailand and Vietnam. ***India:*** Andaman & Nicobar Is., Andhra Pradesh, Arunachal Pradesh, Assam, Bihar, Jammu & Kashmir, Jharkhand, Karnataka, Kerala, Madhya Pradesh, Maharashtra, Meghalaya, Nagaland, Odisha, Sikkim, Tamil Nadu, Telangana, Uttarakhand, Uttar Pradesh and West Bengal. ***Eastern Ghats:*** *AP:* Seshachalam hills and Veligonda hills; *OD:* Mahendragiri hills, Rebana forest and Similipal forest; *TE:* Manannur forest; *TN:* Kolli hills.

Specimens Examined: *AP:* Nichela kanuma (KDP), *BR & BSS* 31066 (SKU); Balapalle (KDP), *JLE* 15744 (MH & CAL); Tirumala (CTR), *BR & BSS*

Distribution: ***World*:** Australia, Bangladesh, Bhutan, Borneo, Celebes, China, India, Java, Malaysia, Myanmar, Nepal, New Guinea, Niue, Ryukyu, Papua, Philippines, Sri Lanka, Sulawesi, Sumatra, Taiwan, Thailand, Timor and Tonga. ***India:*** Andaman & Nicobar Is., Andhra Pradesh, Arunachal Pradesh, Assam, Bihar, Jharkhand, Karnataka, Kerala, Manipur, Meghalaya, Mizoram, Odisha, Sikkim, Tamil Nadu, Telangana, Tripura, Uttar Pradesh and West Bengal. ***Eastern Ghats:*** Commonlly found in Eastern Ghats.

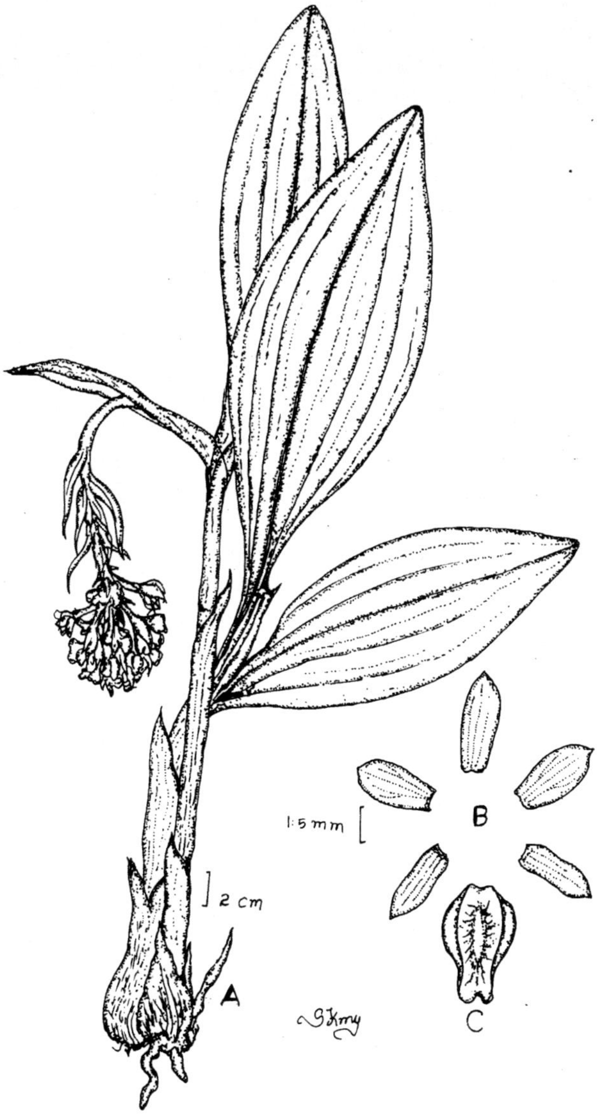

Figure 3. **Geodorum densiflorum** (Lam.) Schltr.
A. Habit; B. Sepals & petals; C. Lip

Specimens Examined: AP: Nagaluty (KNL), *BR* & *BSS* 30570 (SKU); Upper Ahobiilam (KNL), *S.Sunitha* & *AMR* 24405 (SKU); Gundlabrahmeswaram (KNL), *MVS* & *V.Srinivasa Rao* 35165 (SKU); Peddacheruvu (KNL), *K.N.Reddy* 77221 (FRLH); Gundlabramheswaram (KNL), *JLE* 16915 (MH), Chelama (KNL), *JLE* 18044 (MH); Gajulapalli RF (KNL), *JLE* 18044 (MH); way to Diguvametta (KNL), *JLE* 22181 (MH); Caneplot (KNL), *JLE* 25530 (MH); Chakirevula (KDP), *BSS* & *SKB* 30579 (SKU, BSID); Near Lankamalleswaram (KDP), *SRSR* 14636 (SKU); Talakona (CTR), *BSS* & *MVS* 30266 (SKU); Chittoor, *PS* 3402 (PCM); Chinarutla (PKM), *BR* & *KP* 34608 (SKU); Cheedipalem (EG), *GVS* 24519 (MH), Palagadda hill top (EG), *SS* 5583 (AU); Chintapalli towards Narsiptnam (VSKP), *GVS* 30085 (MH). *KA*: B.R.Hills (CRR), *KS* 23287 (SKU). *OD:* Saparai (KP), *SM* 2040 (CAL). *TE*: Jalpenta (MBNR), *BSS* & *KP* 32614 (SKU). *TN:* Javvadhu hills (TVM), *RVS* 43616 (FRLH); Kolli hills (NK), *KMM* & *VAA* 3170, *KMM* 4351, *DIA* 7568, 7605 & 10241, *KMM* & *NV* 18455, *KMM* & *CM* 18620 (RHT); Shevaroy hills, Yercaud, Kiliyur (SLM), *DIA* 2880 (RHT); Shevaroys (SLM) *DIA* 4169, *KMM* 23482 (RHT); Periyakalrayan (SLM), *DIA* 5456, *KMM* 15295 (RHT); Jawalgiri (DMP), *NV* 23975 (RHT).

Geodorum laxiflorum Griff., Calcutta J. Nat. Hist. 5: 356. 1845; Hook.f., Fl. Brit. India 6: 18. 1890; Misra, Orchids Orissa 560. 2004.

Type: India: Assam, *Jenkins s.n.* (not located).

Etymology: The specific epithet refers to laxly-flowered inflorescence of the species.

Terrestrial herbs. Corms in chain, annular, light whitish-brown, compressed, globular to ovoid, *ca* 3.5 × 2 cm; stem stout, terete, *ca* 9 cm long. Leaves 2-3, plicate, elliptic-lanceolate, acute to acuminate, *ca* 25 × 6.5 cm; petiole *ca* 2 cm long. Inflorescence lateral, shorter than the leaves, apex hooked, *ca* 20 cm long; raceme laxly 4-8-flowered. Floral bracts shorter than the ovary, green, oblong-lanceolate, acute, 3-veined, *ca* 7 × 2 mm. Pedicel with ovary greenish-white, *ca* 1 cm long. Flowers white, *ca* 2.6 cm across. Sepals oblong-lanceolate or obovate, acute, 7-veined, *ca* 2.4 × 0.7 cm. Petals elliptic or obovate, acute, 7-veined, *ca* 2 × 1 cm. Lip sessile, cymbiform or square-shaped, base ventricose, emarginated, *ca* 2 × 1.7 cm; epichile margins deflexed, undulate, with two irregular rows of thick warts starting at the base of the hypochile and ending before the apex; hypochile golden brown within, epichile yellow at base and pink at apex. Column white, oblong-obovate, *ca* 5 × 2 mm; foot *ca* 2 mm long. Stigma quadrate. Pollinia 2, oblong-ovoid, 2 × 1.4 mm. Capsules spindle-shaped, *ca* 4 cm long.

Fl. & Fr.: June-August.

Habitat: Rare, it grows under shady conditions in humus rich soils of moist deciduous forest (above 200 m).

Distribution: Endemic to India (Andhra Pradesh, Assam, Chhattisgarh, Gujarat, Jharkhand, Odisha and West Bengal). ***Eastern Ghats:*** *AP:* Nallamala hills; *OD:* Mahendragiri hills.

Specimens Examined: *AP:* Nallamala hills, Rollapenta (PKM), *BR* & *KP* 34616 B & 34617 (SKU). *OD:* Baibali forest (GJM), *SM* 2359 (CAL).

Geodorum recurvum (Roxb.) Alston, Handb. Fl. Ceylon 6 (Suppl.): 276. 1931; Misra, Orchids Orissa 554. 2004; Misra *et al.,* Bull. Bot. Surv. India 50: 137. 2008. *Limodorum recurvum* Roxb., Pl. Corom. 1: 33. t. 39. 1795 *et* Fl. Indica 3: 469. 1832.

Type: India: Andhra Pradesh, Circars, Roxburgh (not located).

Etymology: The specific epithet refers to the recurved nature of inflorescence of the species.

Terrestrial herbs. Corms in chain, greenish-brown, compressed, ovoid, *ca* 4 × 3 cm; stem stout, terete, *ca* 5 cm long. Leaves 2-3, plicate, narrowly elliptic-lanceolate, acute, *ca* 35 × 10 cm; petiole *ca* 2 cm long. Inflorescence lateral, shorter than the leaves, apex hooked, *ca* 25 cm long; raceme densely 10-12-flowered. Floral bracts longer than the ovary, greenish-white, oblong-lanceolate, acuminate, 5-veined, *ca* 2.5 × 0.3 cm. Pedicel with ovary greenish-white, *ca* 8 mm long. Flowers whitish-green, *ca* 2.4 cm across. Sepals oblong-oblanceolate, acute, 3-5-veined; dorsal sepal *ca* 2 × 0.7 cm; laterals *ca* 2.2 × 1 cm. Petals obovate-lanceolate, acute, 5-veined, *ca* 1.8 × 0.8 cm. Lip sessile, cymbiform, undulate, truncate or rounded, *ca* 2.2 × 1.5 cm; disc with two parallel straight calli, with yellow patch beyond the middle and two deeper rose colour sub-orbicular patches. Column whitish-yellow, oblong-ovate, *ca* 7 × 4 mm; foot *ca* 1 mm long. Stigma quadrate. Pollinia 2, oblong-ovoid, 1.6 × 1 mm. Capsules spindle shaped, *ca* 4 cm long.

Fl. & Fr.: May- September.

Habitat: Occasional, it grows under shady condition in humus soils of moist deciduous forests (above 600 m).

Distribution: ***World*:** Bhutan, China, Laos, Myanmar, Sri Lanka and Thailand. ***India:*** Andhra Pradesh, Assam, Jharkhand, Odisha and Telangana. ***Eastern Ghats:*** *AP:* Nallamala hills, Maredumilli forest, Kotapalli forest; *OD:* Commonly distributed Eastern Ghats parts of Odisha; *TE*: Mannanur forest.

Specimens Examined: *AP:* GBM (KNL), *MVS & DV* 35162 (SKU); Opposite Medicinal Plant Conservation Area (EG), *SM* 108; Addatigala (EG), *MSR* 1690 (CAL); Kotapalli (VSKP), *SM* 108A (RPRC). *OD:* Similipal forest-Bhanjabasa (MB), *SM* 833 (CAL); Baibali forest (GJM), *SM* 153 (CAL); Mohana (GJP), *SM* 15 (CAL); Rebana forest (KJ), *G. Panigrahi* 8464 (CAL). *TE:* Saleswaram (MBNR), *BR & KP* 34613 (SKU).

Goodyera R. Br.

Hortus Kew., ed. 2 (W.T. Aiton) 5: 197. 1813.

World: 101/India:19/EG:4

Type: *Goodyera repens* (L.) R. Br. (lectotype)

Etymology: The genus was named in honour of English Botanist, John Goodyer (1592-1664).

Distribution: Europe, Madeira, Mozambique, W. Indian Ocean to Pacific, N. & C. America, Caribbean.

Key to the species

1a. Leaves thick green with reticulated cream-coloured veins .... **G. hispida**

1b. Leaves green colour without coloured veins 2

2a. Sepals hairy on outer surface; petals connate at apex .... **G. thailandica**

2b. Sepals glabrous; petals free 3

3a. Hypochile of lip quadrate when spread, hairy inside and epichile contracted to long ligulate lobe........ **G. fumata**

3b. Lip cymbiform or obscurely 3-lobed........ **G. procera**

Goodyera fumata Thwaites, Enum. Pl. Zeyl. 314. 1861; Hook.f., Fl. Brit. India 6: 111. 1890; Misra, Orchids Orissa 295. 2004.

Type: Sri Lanka: Central Province, *Thwaites* 3668 (holo: K (image!)).

Etymology: The specific epithet refers to the murky colour of flowers.

Inflorescence *ca* 60 cm long; spike densely many-flowered. Floral bracts shorter than the ovary, green, ovate-lanceolate, acuminate, 3-veined, *ca* 2 × 0.6 cm. Pedicel with ovary greenish-yellow, *ca* 2 cm long. Flowers pale white. Sepals acute or obtuse, 3-veined; dorsal ovate, *ca* 4.2 × 2.6 mm; laterals spreading and reflexed, ovate-oblong, *ca* 5 × 2.8 mm. Petals oblong-ovate, mucronate, 1-veined, *ca* 3.2 × 1.8 mm. Lip decurved, ovate-oblong in outline, *ca* 4.5 mm long, 3-lobed near the base; laterals slightly longer or shorter than the midlobe, linear-oblong, obtuse; midlobe linear-lanceolate, obtuse; spur shorter than the ovary, greenish-white, slightly inflated, somewhat clavate towards apex, *ca* 1.6 cm long. Stigmatic lobes 2, broadly ovate-rounded. Pollinia oblong-obovoid, 1.2 × 0.8 mm. Capsules elliptic, *ca* 1.5 cm long.

Fl. & Fr.: August-October.

Habitat: Rare, it grows in high altitude grass lands near margins of semi evergreen and moist deciduous forests (above 800 m).

Distribution: Endemic to Peninsular India (Kerala, Maharashtra, Odisha and Tamil Nadu). ***Eastern Ghats:*** *OD:* Mahendragili hills.

Specimens Examined: *OD:* Mahendragiri (GJM), *VNS* 5775 (MH).

Habenaria crinifera Lindl., Gen. Sp. Orchid. Pl. 323. 1835; Hook.f., Fl. Brit. India 6: 142. 1890; C.E.C. Fisch., Fl. Madras 1471. 1928

Type: Sri Lanka: *Macrae* 36 (holo: K (image!)).

Etymology: The specific epithet refers to the much prolongations nature of lip.

Terrestrial herbs, *ca* 40 cm high. Tubers ovoid or oblong, *ca* 3 × 1.5 cm. Leaves clustered towards base of stem, sessile, oblong-lanceolate, acute, *ca* 15 × 2 cm. Inflorescence *ca* 25 cm long; raceme laxly few-flowered. Floral bracts shorter than the ovary, ovate-acuminate, *ca* 2 cm long. Pedicel with ovary green, *ca* 3 cm long. Flowers white. Dorsal sepal ovate-oblong, obtuse, 3-veined, *ca* 6 × 4 mm; lateral sepals oblong-elliptic, obtuse or rounded, *ca* 8 × 6 mm. petals linear or oblanceolate, acute, 1-veined, *ca* 8 × 1.5 mm. Lip *ca* 2 cm long, clawed, base with horn, 3-lobed; laterals cuneiform, serrate, apex with *ca* 8 mm long filiform prolongation; midlobe flabelliform, 2-lobulate; lobules with mucro in between, serrate, apex with 1 cm long prolongation; spur longer than the ovary, clavate, *ca* 4.5 cm long. Capsules fusiform, *ca* 3 cm long.

Fl. & Fr.: August-November.

Habitat: Rare, it grows in humus rich soils under shady condition of Semi evergreen and moist deciduous forest (above 600 m).

Distribution: ***World***: India and Sri Lanka. ***India***: Karnataka, Kerala, Maharashtra and Tamil Nadu. ***Eastern Ghats:*** *TN:* Kolli hills.

Specimens Examined: *TN:* Kolli hills (NK), *P.G. Diwakar & P. Dwarakan* 97079 (MH).

Habenaria digitata Lindl., Gen. Sp. Orchid. Pl. 307. 1835; Hook.f., Fl. Brit. India 6: 134. 1890; C.E.C. Fisch., Fl. Madras 1469. 1928; Haines, Bot. Bihar & Orissa, 1154. 1924; Misra, Orchids Orissa 179. 2004; Ramakrishna in Pullaiah, Fl. Andhra Pr. 3: 943. 1997; Misra *et al.,* Bull. Bot. Surv. India 50: 137. 2008; Seidenfaden in K.M. Matthew, Fl. Tamilnadu Carnatic 3: 1562. 1983.

Types: Bangladesh or Assam? (India): Wall. Cat. no. 7063 (CAL! & K(image!)).

Etymology: The specific epithet refers to the divided petals and lip (finger shaped) of the species.

Terrestrial herbs, *ca* 60 cm high. Tubers 1-2, elliptic-ovoid, *ca* 2.5 × 1.5 cm; stem stout, terete. Leaves cauline, evenly distributed in upper part of the stem and gradually changing into floral bracts, membranous, sessile, obovate-oblong or ovate or ovate-oblong or elliptic-oblong, acute, *ca* 12 × 5 cm. Inflorescence *ca* 16 cm long; spike densely or laxly few-flowered. Floral bracts as long as or longer than the ovary, pale green, foliaceous, ovate-lanceolate, acuminate, 3-veined, *ca* 3 × 1.4 mm. Ovary subsessile or sessile, green, fusiform, *ca* 2 cm long, beaked. Flowers green with white base, *ca* 2.5 cm across. Dorsal sepal concave, broadly ovate or ovate-oblong, acute, 3-veined, *ca* 9 × 6.5 mm; lateral sepals drooping, falcate, ovate-oblong, acute, *ca* 10 × 6 mm. Petals bipartite nearly to base, lobes unequal; upper lobe falcate, curved, linear, acute, *ca* 1.2 × 0.2 cm; lower lobe lying on the lip, strongly sickle-shaped, filiform, *ca* 6 mm long. Lip fleshy, sessile, *ca* 1.4 cm long, 3-lobes little above the base; laterals distal halves upcurved, linear-filiform, acute, *ca* 1.2 cm long; midlobes as long as the lateral lobes, reflexed, linear, acute; spur shorter or as long as ovary, green, clavate, mouth with erect ligule, *ca* 2 cm long. Stigmatic lobes 2, narrowly oblong. Pollinia 2, obovoid, 2 × 1.2 mm. Capsules ellipsoid, *ca* 2.5 cm long.

Fl. & Fr.: July-October.

Habitat: Rare, it grows under shady condition in humus rich soils of moist deciduous forest (400-800 m).

Distribution: ***World*:** Bhutan, Bangladesh, China, India, Laos, Myanmar, Nepal and Pakistan. ***India:*** Andhra Pradesh, Arunachal Pradesh, Assam, Chhattisgarh, Gujarat, Himachal Pradesh, Jammu & Kashmir, Jharkhand, Karnataka, Kerala, Madhya Pradesh, Maharashtra, Meghalaya, Odisha, Rajasthan, Sikkim, Tamil Nadu, Uttarakhand and West Bengal. ***Eastern Ghats:*** *AP:* Rampa hills, Venkatagiri hills; *OD:* Gopalpur-Gunupur forest; *TN:* Chitteri hills.

Specimens Examined: *AP:* Rajavommangi RF (EG), *GVS* 68640 (MH & CAL); Kappakonda hill (EG), *MSR* 1650 (CAL); Vatigadda (EG), *SS* 827; Venkatagiri hills (NLR), *BSN* 3606 and 3966 (VV); Way to Maredumilli from Rampchodavaram (EG), MM 102681 (BSID). *OD:* Kumdabahl (KH), *SM* 398 (CAL); Pottangi (KP), *HFM* 3943 (DD). *TN*: Chitteri (SLM), *KMM* 20917 (RHT).

Habenaria diphylla (Nimmo) Dalzell, Hooker's J. Bot. Kew Gard. Misc. 2: 262. 1850; Hook.f., Fl. Brit. India 6: 151. 1890; C.E.C. Fisch., Fl. Madras 1471. 1928; Haines, Bot. Bihar & Orissa, 1158. 1924; Misra, Orchids Orissa 205. 2004; Misra *et al.,* Bull. Bot. Surv. India 50: 137. 2008. *Liparis diphyllos* Nimmo, Cat. Pl. Bombay 252. 1839.

Types: India: Concan, *Dalzell s.n.* (K & GH).

Etymology: The specific epithet refers to the two leaves (generally) of the species.

Terrestrial herbs, *ca* 30 cm high. Tubers 1-2, elliptic-obovoid or ellipsoid, *ca* 1.8 × 1 cm; stem slender. Leaves 2-3, lying flat on the ground, sessile, ovate-lanceolate to sub-orbicular, base cordate, apex acute, *ca* 5 × 4 cm. Inflorescence *ca* 30 cm long; raceme laxly many-flowered. Floral bracts shorter than the ovary, greenish-yellow, ovate-lanceolate, acuminate, 1-veined, *ca* 1 × 0.5 cm. Pedicel with ovary greenish-white, *ca* 2.6 cm long, beaked. Flowers *ca* 1.3 cm across. Dorsal sepal green, ovate, acute, 3-veined, *ca* 6 × 4 mm; lateral sepals white, spreading, ovate-lanceolate, acute, *ca* 8 × 4 mm. Petals greenish-white, falcate, linear-oblong, acute or obtuse, 3-vined, *ca* 7 × 1.6 mm. Lip greenish-white, sessile, 3-lobed near the base; laterals spreading, longer than the midlobe, filiform, acute, *ca* 2 cm long; midlobe straight or slightly curved, linear-filiform, acute, *ca* 1 cm long; spur greenish-white, shorter or longer than the ovary, clavate at apex, *ca* 2.8 cm long. Stigmatic lobes 2, deep green, clavate. Pollinia 2, ovoid-oblong, 1.6 × 1 mm. Capsules ellipsoid or fusiform, *ca* 3 cm long.

Fl. & Fr.: August-November.

Habitat: Extremely rare, it grows under shady conditions in humus soils of moist deciduous forest (400-800 m).

Distribution: ***World***: Bangladesh, Bhutan, China, India, Myanmar, Nepal, Philippines and Thailand. ***India***: Andhra Pradesh, Bihar, Chhattisgarh, Jharkhand, Karnataka, Kerala, Maharashtra, Odisha, Sikkim, Tamil Nadu, Uttarakhand and West Bengal. ***Eastern Ghats:*** *AP:* Visakhapatnam forest; *OD:* Rebana forest.

Specimens Examined: *AP:* Araku (VSKP), *NPBK* 10856 (CAL); Anantagiri (VSKP), *V. Seshavataram s.n.* (AU). *OD:* Rebana forest (KJ), *SM* 460 (ORC).

Habenaria furcifera Lindl., Gen. Sp. Orchid. Pl. 319. 1835; Hook.f., Fl. Brit. India 6: 149. 1890; Haines, Bot. Bihar & Orissa, 1157. 1924; Mooney, Suppl. Bot. Bihar & Orissa 207. 1956; Misra, Orchids Orissa 227. 2004; Ramakrishna in Pullaiah, Fl. Andhra Pr. 3: 943. 1997; Misra *et al.,* Bull. Bot. Surv. India 50: 138. 2008.

Type: India: Mussoorie, *Royle s.n.* (holo: K-LINDL).

Etymology: The specific epithet refers to the divided lip of the species.

Terrestrial herbs, *ca* 80 cm high. Tubers 1-3, ovoid-ellipsoid, *ca* 4 × 2 cm. Leaves 4-6, clustered below or above the middle of the stem, sub-sessile, ovate-elliptic or elliptic-obovate or oblanceolate, acuminate, *ca* 8 × 4 cm. Inflorescence *ca* 40 cm long; spike cylindrical, densely many-flowered. Floral bracts shorter or longer than the ovary, green, concave, ovate-lanceolate, acuminate, 1-veined, *ca* 1 × 0.3 cm. Ovary sessile, green, *ca* 1 cm long. Flowers green, *ca* 1.2 cm across. Dorsal sepal concave, broadly ovate, obtuse, 3-veined, *ca* 4 × 2.5 mm; lateral sepals reflexed, falcate, ovate-lanceolate, obtuse, *ca* 4 × 2 mm. Petals falcate, oblong-lanceolate, acute, 1-veined, *ca* 4 × 1.5 mm. Lip sessile, 3-lobed near the base; laterals longer than the midlobe, diverging, linear, obtuse, *ca* 6 × 0.6 mm; midlobe linear-oblong, obtuse, *ca* 4 × 1 mm; spur short or longer than ovary, green, laterally compressed, *ca* 1.4 cm long. Stigmatic lobes 2, elliptic-oblong. Pollinia 2, obovoid, 0.6 × 0.4 mm. Capsules fusiform, *ca* 1.2 cm long.

Fl. & Fr.: August-December.

Habitat: Occasional, it grows under open or shady conditions in humus soils of semi evergreen and moist deciduous forest (above 600 m).

Distribution: ***World*:** Bangladesh, Bhutan, China, India, Laos, Myanmar, Nepal, Pakistan and Thailand. ***India***: Andhra Pradesh, Arunachal Pradesh, Assam, Bihar, Chhattisgarh, Gujarat, Himachal Pradesh, Jammu & Kashmir, Jharkhand, Karnataka, Kerala, Madhya Pradesh, Maharashtra, Meghalaya, Odisha, Punjab, Sikkim, Tamil Nadu, Telangana, Uttar Pradesh, Uttarakhand and West Bengal. ***Eastern Ghats:*** Occasionally found in Eastern Ghats of Andhra Pradesh, Karnataka, Odisha, Telangana.

Specimens Examined: *AP:* GBM to Ramanpenta (KNL), *BR* & *KP* 34634 (SKU); Talkona (CTR), *BR*, *BSS* & *MVS* 32747 (SKU); Chittoor, *PS* 3403 (PCM); Rajavomangi (EG), *SS* 707; Elarthi-Donkarayi (EG), *SS* 5422(AU); Gudem (VSKP), *BR, KP* & *MVS* 34684 (SKU); Near Araku (VSKP), *NPBK* 10856 (MH & CAL); Cherukonda (VSKP), *GVS* 28152 (MH). *OD:* Similipal forest-Bakua (MB), *SM* 1307 (CAL); Rajgangpur, *Jardon* 32 (CAL); Niamgiri (KH), *G.S. Deb* 3109 (DD); Curangi (GJM), *VNS* 5827 (MH). *TE:* Gundam-Mannanur (MBNR), *KP* 6430 (BSID). *TN:* Valasamalai R.F. (TVM), *KRK* 14817 (FRLH).

Habenaria grandifloriformis Blatt. & McCann, J. Bombay Nat. Hist. Soc. 36: 17. 1932; Misra, Orchids Orissa 176. 2004; Seidenfaden in K.M. Matthew, Fl. Tamilnadu Carnatic 3: 1561. 1983. *H. grandiflora* Lindl. *ex* Dalzell & Gibson, Bambay Fl. 267. 1861 (*nom. illeg.*); Hook.f., Fl. Brit. India 6: 136. 1890; C.E.C. Fisch., Fl. Madras 1469. 1928.

Syntypes: India: Maharashtra, Panchgani, on Tableland, *Blatter* P. 20, P. 22, P. 23 (BLAT) and *Sedgewick* 7900 (BLAT).

Etymology: The specific epithet refers to large flowers of the species.

Terrestrial herbs, *ca* 20 cm high. Tubers 1-2, ellipsoid, *ca* 2.5 × 1 cm. Leaves 1-2, lying flat on the ground, sessile, broadly orbicular or ovate-cordate or cordate-rounded or oblong-lanceolate, acute, *ca* 6 × 8 cm. Inflorescence *ca* 18 cm long; raceme laxly 1-3-flowered at apex. Floral bracts much shorter than the ovary, green, broadly ovate or ovate-orbicular, acute, 3-veined, *ca* 1.1 × 0.8 cm. Pedicel with ovary green, *ca* 4 cm long. Flowers fragrant, white. Dorsal sepal concave, broadly ovate, acute or obtuse, 3-veined, *ca* 9 × 6 mm; lateral sepals spreading, ovate-triangular, sub-acute, 5-7-veined, *ca* 11 × 7 mm. Petals bipartite near the base; upper lobe erect, ovate-triangular, acute, *ca* 8 × 7 mm; lower lobe much longer than the upper one, spreading horizontally, filiform or linear, acute, *ca* 15 cm long. Lip *ca* 2.2 cm long, 3-lobed little above the base; laterals spreading, as long as or slightly longer than the midlobe, linear, acute, *ca* 1.8 cm long; midlobe linear-oblong, acute, *ca* 15 × 3 mm; spur whitish-green, cylindrical, dilated towards apex, rounded, *ca* 3 cm long. Stigmatic lobes 2, parallel, oblong. Pollinia 2, ovoid, 2 × 1 mm. Capsules oblong-cylindrical, *ca* 3 cm long.

Fl. & Fr.: June-August.

Habitat: Rare, it grows in open high altitude grass lands (above 900 m).

Distribution: Endemic to Peninsular India (Chhattisgarh, Goa, Gujarat, Karnataka, Kerala, Maharashtra, Odisha and Tamil Nadu). ***Eastern Ghats:*** *OD:* Mahendragiri hills; *TN:* Shevaroy hills.

Specimens Examined: *OD:* Karlapat (KH), *HFM* 3471 (DD); Jeypore estate (KP), *HFM* 3834 (DD); Mahendragiri (GJM), *VNS* 5582 (MH). *TN:* Shevaroy hills, Sanyasimalai (SLM), *KP* 6462 (BSID); Shevaroy hills (SLM), *KMM* 23328 (RHT).

Habenaria heyneana Lindl., Gen. Sp. Orchid. Pl. 320. 1835; Hook.f., Fl. Brit. India 6: 148. 1890; C.E.C. Fisch., Fl. Madras 1471. 1928; Seidenfaden in K.M. Matthew, Fl. Tamilnadu Carnatic 3: 1567. 1983.

Types: India: Wall. Cat. no. 7044 (CAL! & K (image!)).

Etymology: This species was named in honour of Dr. B. Heyne (1770–1819), German botanist, Plant collector from Travancore, SW India and collected type material.

Terrestrial herbs, *ca* 40 cm high. Tubers 1-2, ovoid-oblong, *ca* 4 × 1.5 cm; stem stout, terete. Leaves cauline, along the stem, sessile, narrowly lanceolate or oblong-lanceolate, acute, *ca* 6 × 2 cm. Inflorescence secund or not, *ca* 15 cm long; raceme laxly few-flowered. Floral bracts longer than the ovary, green, oblong-lanceolate, acute, 3-5-veined, *ca* 2 × 4 mm. Pedicel with ovary green, *ca* 2 cm long. Flowers white or greenish-white, turning yellow at age. Dorsal sepal oblong-elliptic or oblong-ovate, acute, 1-3-veined, *ca* 7.5 × 2.8 mm; lateral sepals spreading, oblong-lanceolate, acute, *ca* 7.5 × 2 mm. Petals falcate, linear to oblong-lanceolate, acute, 1-3-veined, *ca* 7 × 2.8 mm. Lip deflexed below the middle, *ca* 8 mm long, 3-lobed below the middle; laterals shorter than the midlobe, falcate, narrowly oblong, obtuse, *ca* 3 mm long; midlobe oblong or oblong-lanceolate, obtuse, *ca* 6 × 1.5 mm; spur shorter than the ovary, green, narrowly cylindrical, *ca* 6 mm long. Stigmatic lobes 2, parallel, clavate. Pollinia 2, globular, 1 × 1 mm. Capsules fusiform, *ca* 2.5 cm long.

Fl. & Fr.: September-December.

Habitat: Rare, it grows in high altitude grass lands and Shola forest (above 800 m).

Distribution: Endemic to Peninsular India (Goa, Karnataka, Kerala, Maharashra and Tamil Nadu). ***Eastern Ghats:*** *TN:* Shevaroy hills.

Specimens Examined: *TN:* Shevaroy temple-grass land (SLM), *KP* 6486 (BSID); Shevaroy hills (SLM), *AVNR* 18265 (MH) & *KMM* & *DIA* 4921 (RHT); Yercaud (SLM), *KMM*, *SJB* & *NRI* 28548 (RHT).

Habenaria hollandiana Santapau, Fl. Purandhar 126. 1957; Ramakrishna in Pullaiah, Fl. Andhra Pr. 3: 944. 1997. *H. indica* C. S. Kumar & Manilal, Taxon 35: 719. 1986 (Illegitimate); Misra *et al.,* Bull. Bot. Surv. India 50: 138. 2008. *H. affinis* Wight, Icon. Pl. Ind. Orient. 5: t. 1707 (Illegitimate); Hook.f., Fl. Brit. India 6: 149. 1890; C.E.C. Fisch., Fl. Madras 1471. 1928.

Type: India: Karnataka, Belgaum, *Law s.n.* (K).

Terrestrial herbs, *ca* 90 cm high. Tubers 1-2, ellipsoid or ovoid. Leaves 3-5, clustered near the middle of the stem, elliptic-lanceolate to oblanceolate, acute to acuminate, *ca* 10 × 3.5 cm. Inflorescence *ca* 30 cm long; raceme densely many-flowered. Floral bracts as long as or longer than ovary, linear-lanceolate, acuminate. Pedicel with ovary green, beaked. Flowers green. Sepals ovate-oblong, *ca* 3.5 mm long. Petals oblong-lanceolate. Lip linear at base then 3-lobed; laterals longer than midlobe, spreading, narrowly linear; midlobe straight, oblong-lanceolate; spur shorter than ovary, clavate, *ca* 8 mm long. Pollinia 2, pyriform. Capsules fusiform, *ca* 2 cm long.

Fl. & Fr.: September-December.

Habitat: Rare, it grows in open or shady conditions in humus rich soils of moist and dry deciduous forests.

Distribution: Endemic to India (Andhra Pradesh, Arunachal Pradesh, Jharkhand, Karnataka, Kerala, Maharashtra, Meghalaya, Nagaland and Tamil Nadu). ***Eastern Ghats:*** *AP:* Nallamala hills, Seshachalam hills, Venkatagiri hills, Elarthi forest (East Godavari) and Vizianagaram forest.

Specimens Examined: *AP:* Elarthi (EG), *SS* 5522; Diguvametta (Nallamala hills) (KNL), *JLE* 22195 (MH); Talakona (CTR), *KP* 6450 (BSID); Venkatagiri hills (NLR), *BSN* 3606 (VV); Udayagiri (NLR), *BSN* 6119 & *ASR* 6678 (VV); Guntupaka (VZN), *MV* 6992; Seshachalam hills-Talakona (CTR), *KP* 6450 (BSID).

Habenaria longicorniculata J. Graham, Cat. Pl. Bombay 202. 1839; Misra, Orchids Orissa 215. 2004; Ramakrishna in Pullaiah, Fl. Andhra Pr. 3: 944. 1997; Misra *et al.,* Bull. Bot. Surv. India 50: 138. 2008; Seidenfaden in K.M. Matthew, Fl. Tamilnadu Carnatic 3: 1564. 1983. *H. longicalcarata* A. Rich., Ann. Sci. Nat., Bot. II, 15: 71. 1841; Hook.f., Fl. Brit. India 6: 141. 1890; C.E.C. Fisch., Fl. Madras 1470. 1928; Haines, Bot. Bihar & Orissa, 1156. 1924; Mooney, Suppl. Bot. Bihar & Orissa 207. 1956.

Syntypes: India: Maharashtra, Khandala, Graham s.n. (K); S. Concan, Law & Stocks *s.n.*(K & M); Tamil Nadu, Pulney hills, Wight *s.n.* (K & CAL).

Etymology: The specific epithet refers to the long spur mouth with long triangular ligule of the species.

Terrestrial herbs, *ca* 1 m high. Tubers 1-2, oblong or ovoid-ellipsoid, *ca* 5 × 2 cm; stem slender, terete. Leaves 3-5, clustered near the base of the stem, oblong to elliptic-lanceolate or oblong-lanceolate or oblong-elliptic, acute, *ca* 10 × 3.8 cm. Inflorescence *ca* 95 cm long; raceme laxly 1-6-flowered at apex. Floral bracts shorter than the ovary, greenish-yellow, convolute, oblong-lanceolate,

Fl. & Fr.: October-December.

Habitat: Rare, it grows under deep shady condition in humus-rich soils of moist deciduous forests (above 400 m).

Distribution: Endemic to Peninsular India (Andhra Pradesh, Kerala, Maharashtra, Odisha, Tamil Nadu and Telangana). ***Eastern Ghats:*** *AP:* Maredumilli forest, Nallamala hills, Seshachalam hills; *OD:* Ghumsur forest and Mahendragiri hills; *TE:* Mannanur forest; *TN:* Tiruvannamalai forest.

Specimens Examined: *AP:* Seshachalam hills-Talakona (CTR), *KP* 6438 (BSID); Way to Matta from GBM (KNL), *BSS, SKB* & *KP* 32351 (SKU); Maredumilli-Rampachodavaram (EG), *SM* 0107; Way to Mareduilli from Rampachodavaram (EG), *MM* 102681 (MH) (labelled as *Habenaria multicaudata*); Gundlabrahmeswaram (KNL) *BSS*, *SKB* & *KP*, 32351(SKU). *OD:* Mohana (GJM), *SM* 122 & 176 (CAL); Bhanjanagar (GJM), *SM* 140 (CAL). *TE:* Amrabad forest reserve (MBNR), *STR* 642 (CAL, K and MH). *TN:* Tiruvannamalai district forest (TVM), *RVS* 37594, 46376, 71312 (FRLH).

Habenaria plantaginea Lindl., Gen. Sp. Orchid. Pl. 323. 1835; Hook.f., Fl. Brit. India 6: 141. 1890; C.E.C. Fisch., Fl. Madras 1470. 1928; Haines, Bot. Bihar & Orissa, 1155. 1924; Mooney, Suppl. Bot. Bihar & Orissa 206. 1956; Misra, Orchids Orissa 198. 2004; Ramakrishna in Pullaiah, Fl. Andhra Pr. 3: 945. 1997; Misra *et al.,* Bull. Bot. Surv. India 50: 138. 2008; Seidenfaden in K.M. Matthew, Fl. Tamilnadu Carnatic 3: 1565. 1983. *Plantaginorchis plantaginea* (Lindl.) Szlach., Richardiana 4: 65. 2004.

Syntypes: India: Travancore, Wall. Cat. no. 7053A (CAL! & K), 7053b & 7053c (K).

Etymology: The specific epithet refers to the European *Plantago* which has similar leaves of the species.

Terrestrial herbs, *ca* 40 cm high. Tubers 1-2, ovoid or ellipsoid, *ca* 4 × 2 cm, covered with tomentum. Leaves 2-6, more or less spreading on the ground, sessile, oblong-elliptic or oblong-lanceolate or obovate-oblong, acute, *ca* 10 × 3 cm. Inflorescence *ca* 40 cm long; spike densely many-flowered. Floral bracts shorter than the ovary, green, ovate-lanceolate, acuminate, 1-3-veined, *ca* 1 × 0.4 cm. Ovary pediceled, green, *ca* 1.5 cm long. Flowers white, *ca* 1.5 cm across. Dorsal sepal concave, broadly ovate, obtuse, 3-5-veined, *ca* 5 × 2.6 mm; lateral sepals spreading, oblong-ovate, obtuse, *ca* 5.5 × 4 mm. Petals linear-oblong, acute, 1-veined, *ca* 5 × 2 mm. Lip clawed, spreading, rhomboid, *ca* 1.5 × 1.2 cm, 3-lobed little below the middle; laterals longer than the midlobe, falcate, semi-ovate, truncate-denticulate, *ca* 8 × 5 mm; midlobe straight, linear-oblong, acute, *ca* 6 × 1.8 mm; spur longer than the ovary, cylindrical, acute, *ca* 3 cm long. Stigmatic lobes 2, clasping, elliptic-oblong. Pollinia 2, sub-globose, 1 × 1 mm. Capsules fusiform, *ca* 1.5 cm long.

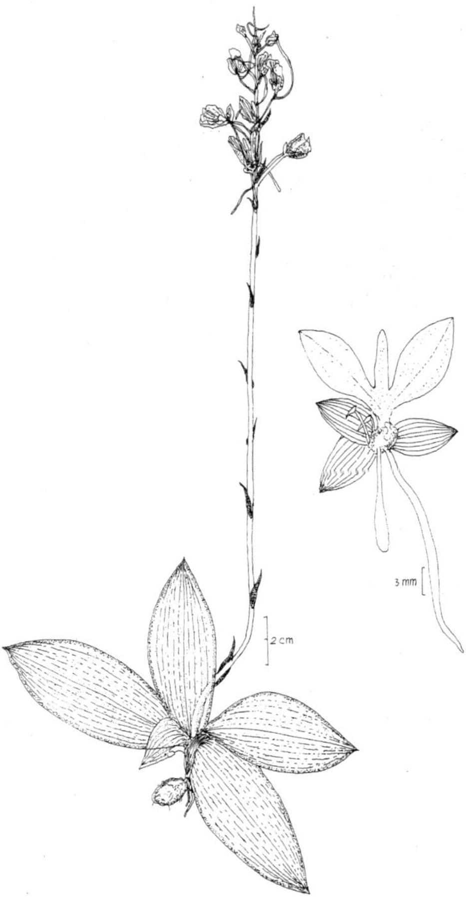

Figure 4. **Habenaria plantaginea** Lindl.

Fl. & Fr.: September-December.

Habitat: Common, it grows in open condition among the grasses of dry and moist deciduous forests.

Distribution: ***World*:** India, Myanmar, Nepal and Sri Lanka. ***India:*** Andhra Pradesh, Bihar, Chhattisgarh, Himachal Pradesh, Jammu & Kashmir, Jharkhand, Karnataka, Kerala, Madhya Pradesh, Maharashtra, Odisha, Punjab, Sikkim, Tamil Nadu, Telangana, Uttarakhand and West Bengal. ***Eastern Ghats:*** Commonly found in all parts of Eastern Ghats.

Specimens Examined: *AP:* Gundlabrahmeswaram (KNL), *BR* & *KP* 34631, 34641 (SKU); Upper Ahobilum (KNL), *SS* & *AMR* 24417 (SKU); Ishtakameswari temple (PKM), *BSS* & *SKB* 31097 (SKU); Way to Maredumilli to Karuru (EG), *KP* 6426 (BSID); Seshachalam hills-Talakona (CTR), *KP* 6449 (BSID); Upper Ahobilum (KNL), *SS* & *AMR* 24417(SKU); Loya, Kondapalli (KSN), *P. Venkanna* 5885 (MH); Addatigala (EG), *KS* 23237 (SKU); Dummakonda RF (EG), *GVS* 68581 (MH & CAL); Rampa Hills (EG), *VNS* 272 (CAL); River Godavari-Sirivaka (EG), *CAB* 4989 (MH); Maredumilli (EG), *MM* 102610 (MH); Way to Ravegudem (EG), *D.C.S. Raju* 472 (CAL): Lododdi reserve (EG), *G.F.F. Foulkes* 63 (CAL); Anantagiri RF (VSKP), *GVS* 10972 (MH); Mear Soleru (VSKP), *GVS* 42715 (MH); Sunkarametta (VSKP), *NPBK* 713 (CAL); Galikonda (VSKP), *GVS* 21644 (MH); Guntupaka (VZN), *MV* 6976 (AU); Udayagiri (NLR), *BSN* 4831 & *ASR* 6675; Venkatagiri (NLR), *BSN* 3606 & *ASR* 6976 (VV). *OD:* Similipal forest-Chahala (MB), *SM* 1393 (CAL); Sambodal (KH), *S.K. Mukerjee* 4375 (CAL); Papadahandi (KP), *T. Das* 29 (DD). *TE:* Gundam-Mannanur (KP), *KP* 6431 (BSID). *TN:* way to Kambukudi (NA), *KS* 7430 (MH); Kottaiyur (NA), *MBV* 825 (MH); Kolli hills (NK), *P.G. Diwakar* & *P. Dwarakan* 97078 (MH); Pulluhalli (SLM), *S. India Flora* 9895 (MH); Pachamalai (TRP), *KMM* 4319, 10780, 17554, 10804, 10838, 10600, 10723 & 19421(RHT); Narthamalai (SLM), *SJB* 30293 (RHT); Periyakalrayan (SLM), *KMM* & *DIA* 5487 (RHT); Puliyancholai (NK), *KMM* & *VKN* 9407 (RHT); Yercaud (SLM), *DIA* 10134 (RHT); Manmalai (SLM), *KMM* 24831 (RHT); Shevaroys (SLM), *KMM*, *SJB* & *NRI* 28002 (RHT); Kalrayans (SLM), *KMM* & *NV* 14005 (RHT); Javalgiri (DMP), *KMM* 24531 (RHT); Thiruvannamalai (TVM), *KMM* 72604 (RHT); Ayyur Forest (DMP), *G. Balanathan* & *P.S. Udayan* 25631 (FRLH).

Habenaria rariflora A. Rich., Ann. Sci. Nat., Bot. II, 15: 70. 1841; Hook.f., Fl. Brit. India 6: 136. 1890; C.E.C. Fisch., Fl. Madras 1469. 1928; Ramakrishna in Pullaiah, Fl. Andhra Pr. 3: 946. 1997; Misra *et al.*, Bull. Bot. Surv. India 50: 138. 2008; Seidenfaden in K.M. Matthew, Fl. Tamilnadu Carnatic 3: 1562. 1983.

Type: India: Tamil Nadu, Nilgiri hills, Coonoor, *Perrottet s.n.* (P).

Etymology: The specific epithet refers to limited number of flowers of the species.

Liparis espeevijii S. Misra, Orchids Orissa: 370. 2004. *L. vestita* subsp. *seidenfadenii* S. Misra, Nordic J. Bot. 6: 26. 1986.

Type: India: Odisha, Kendujhar district, Rebana forest, *SM* 753 (holo:CAL!)

Etymology: This species is dedicated to Prof. S.P. Vij, for his outstanding contribution to Indian orchidology.

Epiphytic herbs. Rhizome short, woody; pseudobulb elongated. Leaves 2, oblong-lanceolate, apex acute, base narrowed into petiole, *ca* 25 × 2 cm. Raceme densely many-flowered. Floral bracts longer than the ovary, subulate, 1-veined, *ca* 4 mm long. Pedicel with ovary *ca* 3.5 mm long. Sepals pale yellow, ovate-lanceolate, acute or obtuse, *ca* 2.5 × 1 mm. Petals yellowish-white, linear-oblong, acute, margins revolute. Lip pale yellow, *ca* 2 mm long, ovate, margins undulate, apex truncate and deeply emarginated.

Fl.: April-May.

Habitat: Very rare, it gows in semi-ever green forest (about 600 m)

Distribution: Endemic to Odisha. ***Eastern Ghats:*** *OD:* Rebana forest.

Specimens Examined: *OD:* Rebana forest-Kendughat (KJ), *SM* 753 (CAL).

Note: This species differs from *Liparis vestita* Rchb.f. by having 5-veined sepals, 3-veined petals, ovate, apex deeply emarginate lip and stigma with large petaloid wings on either side.

Liparis nervosa (Thunb.) Lindl., Gen. Sp. Orchid. Pl. 26. 1830; C.E.C. Fisch., Fl. Madras 1410. 1928; Mooney, Suppl. Bot. Bihar & Orissa 210. 1956; Misra, Orchids Orissa 353. 2004; Misra *et al.,* Bull. Bot. Surv. India 50: 139. 2008. *Ophrys nervosa* Thunb., Syst. Veg. ed. 14. 814. 1784. *Liparis bituberculata* (Hook.) Lindl., Bot. Reg. 11: t. 882. 1825.

Type: Japan: *Thunberg s.n.* (holo:UPS-THUNB).

Etymology: The specific epithet refers to the prominently veined leaves.

Terrestrial herbs. Pseudobulbs green, fusiform, *ca* 1.5 cm long. Leaves 3-5, plicate, ovate-lanceolate or elliptic-lanceolate, margins undulate, apex acuminate, *ca* 10 × 4 cm; petiole *ca* 1 cm long. Inflorescence *ca* 35 cm long, angular; raceme laxly many-flowered. Floral bracts shorter than the ovary, green, deflexed or not, ovate-triangular, acute, 1-veined, *ca* 4 × 3 mm. Pedicel with ovary greenish-maroon or pale maroon, *ca* 9 mm long. Flowers greenish or yellowish-maroon, *ca* 1 mm across. Dorsal sepals linear-oblong, obtuse, 3-5-veined, *ca* 10 × 2 mm; lateral sepals spreading, twisted, oblong-ovate, obtuse, *ca* 8 × 3 mm. Petals drooping or spreading, linear, margins deflexed, obtuse, 1-veined, *ca* 10 × 1 mm. Lip sessile, decurved about the middle, obovate or wedge-shaped, retuse, *ca* 8 × 4 mm; base with rectangular callus and two tubercles on top, with small projection at center. Column yellowish-cream, winged, oblong, *ca* 2.5 × 1.4 mm. Pollinia 4, oblong-ovate, 0.6 × 0.4 mm. Capsules ellipsoid, *ca* 2 cm long.

Fl. & Fr.: June-August.

Habitat: Rare, it grows along perennial streams under shady conditions in humus rich soils of evergreen, semi evergreen and moist deciduous forests (500-1000 m).

Distribution: ***World*:** Bhutan, China, India, Japan, Java, Laos, Malaysia, Moluccas, Myanmar, Nepal, Sri Lanka, Sumatra and Thailand. ***India***: Andhra Pradesh, Arunachal Pradesh, Chhattisgarh, Himachal Pradesh, Jharkhand, Karnataka, Kerala, Maharashtra, Meghalaya, Odisha, Sikkim, Tamil Nadu, Uttarakhand and West Bengal. ***Eastern Ghats:*** *AP:* Seshachalam hills and Kotapalli forest; *OD:* Rebana forest and Similipal forest.

Specimens Examined: *AP:* Way to Rudragal from Talkona top (KDP), *BS & MVS* 32761 (SKU); Kotapalli (VSKP), *SM* 110 (RPRC). *OD:* Similipal forest-Meghasini hills (MB), *SM* 103 (CAL); Similipal forest-Jenabila (MB), *SM* 843 (CAL).

Liparis odorata (Willd.) Lindl., Gen. Sp. Orchid. Pl. 26. 1830. *Malaxis odorata* Willd., Sp. Pl. 4: 91. 1805. *Liparis paradoxa* (Lindl.) Rchb.f., Ann. Bot. Syst. 6: 218. 1861; Hook.f., Fl. Brit. India 5: 697. 1888; Misra, Orchids Orissa 357. 2004; Misra *et al.,* Bull. Bot. Surv. India 50: 139. 2008. *Empusa paradoxa* Lindl., Bot. Reg. 10: t. 825. 1824.

Type: India: Malabar, Rheede, Hort. Malab. (Kattou ponnam marawara) 12: t. 28. 1693.

Etymology: The specific epithet refers to the fragrant flowers of the species.

Terrestrial herbs. Corms broadly ovate-conical, *ca* 3 × 2 cm; stem elongated. Leaves 3-4, plicate, sessile, elliptic-lanceolate, acute or acuminate, *ca* 10 × 4 cm. Inflorescence *ca* 20 cm long, angular; raceme laxly many-flowered. Floral bracts shorter than the ovary, green, deflexed, linear-lanceolate, acuminate, 1-veined, *ca* 6 × 2 mm. Pedicel with ovary yellowish-maroon or green, *ca* 1.4 cm long. Flowers greenish-yellow or pale yellow with maroon tinge, *ca* 8 mm across. Dorsal sepal deflexed, oblong-lanceolate, obtuse, 5-veined, *ca* 6.5 × 1.8 mm; lateral sepals lying below the lip, their tips touching each other or deflexed, falcate, ovate-lanceolate, obtuse, *ca* 6 × 2.4 mm. Petals deflexed, linear, margins revolute, apex obtuse, 3-veined, *ca* 6.2 × 1 mm. Lip sessile, deflexed at middle, narrowly obovate or obovate-oblong or sub-quadrate, emarginated, apical margin crenate, base cordate, 3-veined, *ca* 5 × 5 mm; callus at base, two tubercled on top with deep sinus in between. Column whitish-yellow, winged, oblong, 5 × 1.2 mm. Pollinia 4, ovoid, 0.6 × 0.2 mm. Capsules ellipsoid, *ca* 2 cm long.

Fl. & Fr.: July-August.

Habitat: Occasional, it grows under shady condition in humus rich soils of moist deciduous forest and meadows (400-1100 m).

Distribution: ***World*:** Bangladesh, Bhutan, Cambodia, China, India, Japan, Java, Laos, Myanmar, Nepal, Sri Lanka, Taiwan, Thailand and Vietnam. ***India:*** Andhra Pradesh, Arunachal Pradesh, Assam, Bihar, Chhattisgarh, Himachal Pradesh, Jharkhand, Karnataka, Kerala, Manipur, Meghalaya, Mizoram, Nagaland, Odisha, Sikkim, Tamil Nadu, Telangana, Uttarakhand and West Bengal. ***Eastern Ghats:*** *AP:* Nallamala hills and Seshachalam hills; *OD:* Brahmanigam forest, Gopalpur-Gunupur forest, Govindapalli-Gupteswara forest, Mahendragiri hills, Rebana forest and Similipal forest; *TE*: Nallamalais.

Specimens Examined: *AP:* Rollapenta (KNL), *BR* & *KP* 34671 (SKU); Talakona top (KDP), *BR* & *KP* 34642; Yetakayalamandapam (CTR), *SM* 0303 (RPRC). *OD:* Brahmanigam (KDM), *SM* 163 (CAL); Dudkaguda (KH), *SM* 405 (CAL). *TE:* Sri Sailam Dam site (MBNR), *L.J.G. Van der Maesen* 3292 (CAL).

Liparis viridiflora (Blume) Lindl., Gen. Sp. Orchid. Pl. 31. 1830; Hook.f., Fl. Brit. India 5: 704. 1890; C.E.C. Fisch., Fl. Madras 1411. 1928; Misra, Orchids Orissa 368. 2004; Seidenfaden in K.M. Matthew, Fl. Tamilnadu Carnatic 3: 1585. 1983; Ramakrishna in Pullaiah, Fl. Andhra Pr. 3: 948. 1997; Misra *et al.,* Bull. Bot. Surv. India 50: 139. 2008. *Malaxis viridiflora*

ca 1 cm across. Sepals linear-oblong, margins recurved, apex obtuse, 3-veined; dorsal *ca* 6 × 2 mm; laterals lying below the lip, *ca* 5.6 × 2.2 mm. Petals linear-oblanceolate or linear, margins recurved, apex obtuse, 1-veined, *ca* 6 × 0.8 mm. Lip recurved, suborbicular-obovate, crenulate, 5-veined, *ca* 4.6 × 4.5 mm; base with two tubercles. Column white, winged, *ca* 3 × 1.4 mm. Pollinia 4, pyriform, 0.6 mm.

Fl.: October-November.

Habitat: Extremely rare, it grows under shady conditions of evergreen forests (above 800 m).

Distribution: ***World:*** India & Sri Lanka. ***India:*** Kerala and Tamil Nadu. ***Eastern Ghats:*** *TN:* Shevaroy hills.

Specimens Examined: *TN*: Shevaroy hills (SLM), *DIA* 4071 & 4161, *KMM* 25353 & 25797, *NV* & *TSJ* 22433 (RHT).

Liparis wightiana Thwaites, Enum. Pl. Zeyl. 295. 1861; Hook.f., Fl. Brit. India 5: 695. 1890; C.E.C. Fisch., Fl. Madras 1410. 1928; Seidenfaden in K.M. Matthew, Fl. Tamilnadu Carnatic 3: 1584. 1983.

Types: Sri Lanka: *Thwaites* 3179 (P & BM (images!)).

Etymology: This species was named in honour of well known taxonomist, R. Wight (1796-1872), who collected plants from Peninsular India.

Terrestrial herbs. Leaves broad ovate. Flowers small, in racemes. Dorsal sepal ovate-lanceolate; lateral sepals under the lip, oblong-ovate. Petals longer than the sepals, linear. Lip flat, orbicular or orbicular-oblong, calli at base.

Fl. & Fr.: July-September.

Habitat: Extremely rare; this species grows in high altitude grass lands under shaded by large rocks (above 1000 m).

Distribution: India, Sumatra and Sri Lanka. ***India:*** Karnataka, Kerala and Tamil Nadu. ***Eastern Ghats:*** *TN:* Shevaroy hills.

Specimens Examined: TN: Shevaroy hills, Marapalem, Yercaud (SLM), *AVNR* 26940 (MH) and *R. Ansari* 47795 (MH).

Excluded species

Liparis resupinata Ridl. Was reported by Reddy *et al.* (2005) from Eastern Ghats of Odisha and without any herbarium specimens to support this finding. Futher explorations are needed for confirming its distribution in Eastern Ghats.

Luisia Gaudich.

Voy. Uranie 426. 1829.

World: 39/India:19/EG:5

Type: *Luisia teretifolia* Gaudich.

Etymology: The genus was named in honour of Spanish botanist, Don Luis de Torres (1565-1607).

Distribution: Tropical and Subtropical Asia to West Pacific.

Key to the species

1a. Lip 3-lobed... 2

1b. Lip not 3-lobed 3

2a. Sepals and petals similar; disk of lip 3-ridged... ....**L. abrahamii**

2b. Sepals and petals dissimilar; disk of lip 5-ridged .... **L. birchea**

3a. Lip dividing hypochile and epichile line indistinct and epichile rhomboid .... **L. brachystachys**

3b. Lip dividing hypochile and epichile line distinct and epichile not as above 4

4a. Epichile of lip more than 5 mm broad and with 6-10 streaks .... **L. trichorrhiza**

4b.Epichile of lip less than 5 mm broad and with or without streaks .... **L. zeylanica**

Luisia abrahamii Vatsala, Introd. Orchids 489. 1981.

Type: India: Kerala, Trivandrum, *Vatsala* 207 (holo: Herbarium of Botany, University of Kerala).

Etymology: This species was named honour to A. Abraham, one of the well-known orchidologist from South India.

Distribution: ***World:*** Bangladesh, Bhutan, China, India, Laos, Myanmar, Thailand and Vietnam. ***India:*** Andaman and Nicobar Is., Arunachal Pradesh, Assam, Meghalaya, Odisha, Sikkim, Uttarakhand and West Bengal. ***Eastern Ghats:*** *OD:* Similipal forest.

Specimens Examined: *OD:* Similipal forest-Raivangpur-Jashipur (MB), *SM* 80 (CAL).

Luisia trichorrhiza (Hook.) Blume, Mus. Bot. 1: 63. 1849; Hook.f., Fl. Brit. India 6: 23. 1890; Haines, Bot. Bihar & Orissa, 1177. 1924; Mooney, Suppl. Bot. Bihar & Orissa 214. 1956; Misra, Orchids Orissa 669. 2004; Misra *et al.,* Bull. Bot. Surv. India 50: 140. 2008. *Vanda trichorhiza* Hook., Exot. Fl. 1: t. 72. 1823.

Lectotype: Nepal: Hooker, Exotic Flora t. 72. 1825; designated by Seidenfaden, 1971.

Etymology: The specific epithet refers to the hairy roots of the species.

Epiphytic herbs. Stems stout, terete, woody. Leaves pale green with purple spots or not, fleshy, sessile, terete, *ca* 12 × 0.4 cm. Inflorescence leaf-opposed, condensed, *ca* 1.6 cm long; raceme laxly 4-6-flowered. Floral bracts green, broadly triangular, mucronate, *ca* 2 × 1.8 mm. Pedicel with ovary greenish-yellow with pink spots, *ca* 8 mm long. Flowers *ca* 1 cm across. Sepals and petals greenish-yellow with purple dots or not, apex incurved and obtuse, 5-veined. Dorsal sepal concave, *ca* 6.5 × 3 mm; lateral sepals cymbiform, keeled, oblong-ovate, *ca* 7 × 3.5 mm. Petals oblong-ovate, *ca* 7 × 2 mm. Lip dark purple, fleshy, sessile, *ca* 8 mm long; hypochile deeply concave, sub-quadrate, *ca* 2.2 mm long;

epichile separated from the hypochile by narrow furrow, spreading, ridged, broadly cordate-ovate, obtuse, *ca* 5 × 5 mm. Column yellowish-purple, oblong, *ca* 2.8 × 2 mm. Stigma subquadrate. Pollinia 2, ovoid, 1.2 × 1mm. Capsules clavate or cylindric, *ca* 4 cm long.

Fl. & Fr.: March-July.

Habitat: Occasional, it grows in moist deciduous forest (500-800 m).

Distribution: ***World*:** Bhutan, India, Myanmar, Nepal, Sri Lanka and Thailand. ***India:*** Andhra Pradesh, Arunachal Pradesh, Jharkhand, Karnataka, Meghalaya, Odisha, Sikkim, Uttarakhand and West Bengal. ***Eastern Ghats:*** *AP:* Maredumulli forest and Chintapalli forest; *KA:* BRT hills; *OD:* Brahmanigam, Paniganda, Daringbadi and Kirikuti forests, Chandrapur forest, Gopalpur-Gunupur forest, Govindapali-Gupteswara forest, Kapilas hills, Mahendragiri hills, Rebana forest and Similipal forest.

Specimens Examined: *AP:* Vejuvada (EG), *SM* 90 (RPRC); Chintapalli (VSKP), *SM* 90a (RPRC). *OD:* Similipal forest-UBK (MB), *SM* 1076 (CAL); Mahendragiri hills (GJP), *SM* 1714 (CAL); Kakiregumma (KP), *SM* 22 (CAL).

Luisia zeylanica Lindl., Fol. Orchid. 4: 3. 1853; Misra, Orchids Orissa 665. 2004; Seidenfaden in K.M. Matthew, Fl. Tamilnadu Carnatic 3: 1557. 1983; Ramakrishna in Pullaiah, Fl. Andhra Pr. 3: 949. 1997; Misra *et al.*, Bull. Bot. Surv. India 50: 140. 2008. *L. teretifolia* auct. non Gaudich., Hook.f., Fl. Brit. India 6: 22. 1890; C.E.C. Fisch., Fl. Madras 1438. 1928.

Type: Sri Lanka (Ceylon): *Macrae* 50 (holo: K-LINDL (image!)).

Etymology: This species was named after the type collection country, Zeylan (Old name of Sri Lanka).

Epiphytic herbs. Stems erect or pendulous, woody. Leaves dark green spotted with purple spots, fleshy, sessile, slightly compressed, terete, *ca* 14 × 0.5 cm. Inflorescence extra-axillary, condescend, *ca* 1 cm long; raceme densely 2-6-flowered. Floral bracts pale green, triangular, margins dentate, apex acute, *ca* 1.8 × 2 mm. Pedicel with ovary pale green with purple markings, *ca* 6 mm long. Flowers *ca* 8 mm across. Sepals pale green with maroon dots, incurved, oblong-ovate, obtuse, 1-3-veined, *ca* 4.5 × 2.5 mm. Petals falcate, linear-oblong, *ca* 5 × 1.4 mm. Lip pale yellowish-maroon or dark maroon, fleshy, sessile, *ca* 6 × 4 mm; hypochile concave, sub-quadrate; epichile separated from the hypochile by narrow furrow, deflexed, cordate-ovate or cordate-triangular, obtuse. Column maroon-yellow, oblong, *ca* 3 × 1.4 mm. Stigma oblong-orbicular. Pollinia 2, sub-globose, 0.8 × 0.8 mm. Capsules narrowly fusiform, *ca* 4 cm long.

Fl. & Fr.: February-May.

Habitat: Occasional in moist deciduous forests (500-1000 m).

Distribution: ***World*:** Bhutan, India, Nepal and Sri Lanka. ***India:*** Andaman and Nicobar Is., Andhra Pradesh, Arunachal Pradesh, Assam, Karnataka, Kerala, Madhya Pradesh, Maharashtra, Meghalaya, Odisha, Sikkim, Tamil Nadu, Telangana, Uttarakhand and West Bengal. ***Eastern Ghats:*** *AP:* Gudem valley, Maredumilli forest, Madugula and Minumuluru forests; *KA:* BR hills; *OD:* Mahendragiri hills and Similipal forest; *TE*: Mannanur forest; *TN:* Shevaroy hills.

Specimens Examined: *AP:* Rollapenta (PKM), *BR* & *KP* 34623 (SKU); Galikonda (VSKP), *GVS* 19659 (MH); G. Madugula (VSKP), *GVS* 47288 (MH & CAL); Gudiem valley (VSKP), *VNS* 495 (MH); Minumuluru towards Ontally (VSKP), *GVS* 30029 (MH); Minumuluru (VSKP), *SM* 0101 (RPRC). *KA*: BR hills (CRN), *MVS* & *V.S. Rao* 33756 (SKU). *OD:* Similipal forest-Gudugudia (MB), *SM* 1741 (CAL); Mahendragiri hills (GJP), *SM* 149 (CAL). *TE*: Jalapenta (MBNR), *BSS, SKB* & *KP* 32619 (SKU). *TN:* Shevaroy hills (SLM), *AVNR* 23059 (MH); Shevaroy hills, Yercaud (SLM), *KMM* 4334, *KMM* & *DIA* 2731, *DIA* 7435, *AM* 13583 (RHT); Kolli hills (NK), *DIA* 7609, *KMM* 62716 (RHT); Kolli hills (SLM), *MBV* 17211 (FRLH); Periyakalrayan (SLM), *KMM* 15275, *DIA* 3654 (RHT); Chinnakalrayan (SLM), *KMM* 6839 (RHT); Bodai hills (SLM), *PP* & *CM* 22130 (RHT).

Doubtful species

Luisia filiformis Hook.f. and *L. primulina* E.C.Parish & Rchb.f. both the species were reported by Reddy *et al.* (2005) from Eastern Ghats of Odisha. We have seen no herbarium material from the Odisha and other parts of the Eastern Ghats. Further field studies on these two species are required before confirming its distribution in Eastern Ghats.

Malaxis Sol. *ex* Sw.

Prodr. Veg. Ind. Occ. 119. 1788.

World:163/India: 4/EG: 2

Type: *Malaxis spicata* Sw. (lectotype)

Etymology: The generic name is derived from Greek word, *malaxos* (soften) refers to the soft textured on the leaves.

Distribution: Cosmopolitan.

Key to the species

1a. Flowers pink; front margin of lip with more than 12 pectinate teeth ..... **M. intermedia**

1b. Flowers green or yellowish-purple; front margin of lip with less than 12 pectinate teeth… .... **M. versicolor**

Malaxis intermedia (A. Rich.) Seidenf., Bot. Tidsskr. 73: 99. 1978. *Liparis intermedia* A. Rich., Ann. Sci. Nat., Bot. II, 15: 17. 1841. *Microstylis stocksii* Hook.f., Hooker's Icon. Pl. 19: t. 1833. 1889; Hook.f., Fl. Brit. India 5. 1890; C.E.C. Fisch., Fl. Madras 1408. 1928.

Type: India: Tamil Nadu, Nilgiri hills, *Perrottet s.n.* (G (image!))

Terrestrial herbs. Leaves 2-3, ovate or lanceolate, *ca* 6 mm long. Floral bracts deflexed, lanceolate. Flowers pink, in racemes. Sepals elliptic-oblong; petals linear. Lip reflexed, flabellate, deeply pectinate, segments more than 12.

Fl. & Fr.: August-September.

Habitat: Extremely rare, it grows in humus rich soils of evergreen forest (above 1000 m).

Distribution: Endemic to South India (Karnataka, Kerala and Tamil Nadu). ***Eastern Ghats:*** *KA:* BR hills.

Specimens Examined: *KA:* BR hills (CRN), *E. Barnes s.n.* (DD).

Malaxis versicolor (Lindl.) Abeyw., Ceylon J. Sci., Biol. Sci. 2: 247. 1959. *Microstylis versicolor* Lindl., Gen. Sp. Orchid. Pl. 21. 1830; C.E.C. Fisch., Fl. Madras 3: 1408. 1928; Haines, Bot. Bihar & Orissa, 1165. 1924. *Seidenfia versicolor* (Lindl.) Marg. & Szlach., Polish Bot. J. 46: 56. 2001; Misra, Orchids Orissa 387. 2004; Misra *et al.,* Bull. Bot. Surv. India 50: 143. 2008. *Seidenfia rheedei* auct. non. Szlach., Orchids Orissa 387. 2004. *Malaxis rheedii* Sw., Kongl. Vetensk. Acad. Nya Handl. 21: 235. 1800; Seidenfaden in K.M. Matthew, Fl. Tamilnadu Carnatic 3: 1582. 1983.

Syntypes: India: India Orientalis, *Hayne s.n.* and Sri Lanka: *Macrae s.n.* (not located).

Etymology: The specific epithet refers to the variously coloured flowers of the species.

Terrestrial herbs. Pseudobulbs annular, elongated, *ca* 10 × 0.8 cm, nodded, fully covered with sheaths. Leaves 3-4, plicate, broadly ovate-elliptic, undulate or sub-entire, acuminate, *ca* 11 × 6 cm; petioles *ca* 2 cm long. Inflorescence *ca* 30 cm long; raceme densely many-flowered. Floral bracts green, deflexed, ovate-lanceolate, acuminate, 1-veined, *ca* 3 × 1 mm. Pedicel with ovary yellowish-purple or green, *ca* 4 mm long. Flowers non-resupinate, fragrant, green or yellowish-purple, *ca* 3 mm across. Sepals and petals margins revolute, apex obtuse, 1-veined; dorsal oblong or oblong-lanceolate, *ca* 4 × 1.2 mm; laterals falcate, broadly ovate, *ca* 3.8 × 1.8 mm; petals narrowly linear or linear-lanceolate, *ca* 3.8 × 0.8 mm. Lip reflexed, sessile, reniform or fan-shaped, base concave, front margin broadly truncate with pectinately toothed (teeth about 12), *ca* 2 × 3 mm. Column greenish-yellow, oblong-ovate, 1.4 × 0.8 mm, with arms. Stigma obovate-oblong. Pollinia 4, pyriform, 0.4 × 0.2 mm. Capsules obovate-oblong, *ca* 5 mm long.

Fl. & Fr.: August-October.

Habitat: Rare, it is growing under dense shady condition in humus rich soils of semi evergreen and moist deciduous forests (500-1000 m).

Distribution: ***World*:** China, India, Thailand and Sri Lanka. ***India:*** Andhra Pradesh, Chhattisgarh, Jharkhand, Karnataka, Kerala, Maharashtra, Meghalaya, Odisha and Tamil Nadu. ***Eastern Ghats:*** *AP:* Seshachalam hills and Araku forest; *OD:* Kapilas hills, Mahendragiri hills, Rebana forest and Similipal forest; *TN:* Kolli hills.

Specimens Examined: *AP:* Talkona top (KDP), *BR, BSS* & *MVS* 30441 (SKU), *BR* & *KP* 34674 (SKU); Talakona (CTR), *BR, KP* & *DV* 41628 (SKU); Sunkarimetta (VSKP), *NPBK* 10917 (CAL & MH); Araku valley (VSKP), *DDSR* 21405 (MH). *OD:* Panchpatmali hills (KP), *SM* 234 (CAL); Jeypore estate (KP), *HFM* 3850 (DD); Singaraja hills (GJM), *SM* 202 (CAL); Mahendragiri (GJM), *VNS* 5679 (MH); Kapilas hills (DK), *SM* 340 (CAL). *TN:* Shevaroy hills, Kakka sholai, Yercaud (SLM), *KS* 6597 (MH); Shevaroys bauxite mines (SLM), *AVNR* 18258 (MH); Kollimalais (TRP), *S. India Flora* 11416 (MH); Kolli hills (NK), *KRK* & *RVS* 104369 (FRLH) & *B.Velmurugan* & *RVS* 30292 (FRLH)

Micropera Lindl.

Edwards's Bot. Reg. 18: t. 1522. 1832.

World:100/India:4/EG:1

Type: *Micropera pallida* (Roxb.) Lindl.

Etymology: The generic name is derived from Greek words, *micro* (small) and *pera* (pocket), refers to the lip with a small pocket-like sac.

Micropera pallida (Roxb.) Lindl., Edwards's Bot. Reg. 18: t. 1522. 1832; Misra, Orchids Orissa 610. 2004. *Aerides pallida* Roxb., Fl. Ind. ed. 1832 3: 475. 1832. *Sarcochilus roxburghii* Hook.f., Fl. Brit. India 6: 36. 1890.

Iconotype: Bangladesh: Chittagong, *Roxburgh* Icon. 2349.

Etymology: The specific epithet refers to the colour of the flowers.

Epiphytic scandent herbs. Stem terete, rooted along the stem. Leaves coriaceous, sessile, linear-oblong, *ca* 8 × 1 cm. Racemes as long as or slightly longer than the leaves, densely flowered. Floral bracts broad-triangular, persistent. Pedicel with ovary *ca* 6 mm long. Flowers light yellow. Sepals and petals oblong-lanceolate, obtuse, 3-5-veined; lateral sepals basally connate, *ca* 6 mm long; petals short, *ca* 3.5 mm long. Lip calceolate, porrect, laterally compressed, *ca* 7 mm long, 3-lobed; laterals subquadrate-rounded; midlobe shorter than the laterals, sub-quadrate or triangular, apex with three horns; with deeply bifurcate callus; spur large, with septum. Capsules *ca* 5 cm long.

Fl. & Fr.: June-January.

Habitat: Rare, it is growing in moist deciduous and semi-evergreen forests (above 400 m).

Distribution: ***World:*** Bangladesh, Cambodia, India, Laos, Malaysia, Myanmar, Sumatra, Thailand and Vietnam. ***India:*** Meghalaya, Odisha and Tripura. ***Eastern Ghats:*** *OD:* Rebana forest and Similipal forest.

Specimens Examined: *OD:* Rebana forest-Kendughati (KJ), *SM* 747 (CAL); Similipal forest-Andharojodi (MB), *SM* 1264 (CAL).

Nervilia Comm. *ex* Gaudich.

Voy. Uranie, Bot. 421. t. 35. 1829 (*nom. cons.*).

World:100/India:15/EG:6

Type: *Nervilia aragoana* Gaudich. (*nom. cons.)*

Etymology: The generic name is derived from Latin word, *nervus* (veins), refers to prominent veins on leaves.

Distribution: Africa to Pacific.

Key to the species

1a. Leaf flat on ground and hairy 2

1b. Leaves glabrous, not prostrate 3

2a. Leaf upper surface light green; solitary flower........ **N. crociformis**

2b. Leaf upper surface greenish-maroon or brownish-green; racemes 2-flowered...... .... **N. plicata**

3a. Leaf shortly petioled; solitary flower........ 4

3b. Leaf long petioled; racemes more than 2-flowered 5

4a. Leaf upper face with purple streaks........ **N. infundibulifolia**

4b. Leaf upper face without purple streaks...... ....**N. falcata**

5a. Sepals and petals light green with purple veins...**N. aragoana**

5b. Sepals and petals pink with white veins..**N. gammieana**

Nervilia aragoana Gaudich., Voy. Uranie 422. 1829; C.E.C. Fisch., Fl. Madras 1459. 1928; Misra, Orchids Orissa 259. 2004; Ramakrishna in Pullaiah, Fl. Andhra Pr. 3: 950. 1997; Misra *et al.,* Bull. Bot. Surv. India 50: 140. 2008. *Pogonia flabelliformis* Lindl., Gen. Sp. Orchid. Pl. 415. 1840 (*nom. nud.*); Hook.f., Fl. Brit. India 6: 121. 1890; Haines, Bot. Bihar & Orissa, 1163. 1924. *Nervilia scottii* (Rchb.f.) Schltr., Bot. Jahrb. Syst. 45: 404. 1911; Misra, Orchids Orissa 255. 2004. *Pogonia scottii* Rchb.f., Flora 55: 276. 1872; Hook.f., Fl. Brit. India 6: 120. 1890.

Type: Pacific Is.: Matiannis, *Guaudichaud-Beaupre s.n.* (holo: P).

Etymology: This species is named in honour of Jacques Arago (Artist) who was the artist of Freycinet and Gaudichaud-Beaupre.

Terrestrial hysteranthous herbs. Corm globose or sub-globose, *ca* 1.5 × 1.5 cm; stolons 2, terete, *ca* 10 cm long. Leaf above the ground, plicate, glabrous, orbicular or sub-reniform, base cordate, margins undulate and coarsely dentate, apex acuminate, *ca* 8 × 8 cm; petiole slender, *ca* 15 cm long. Inflorescence angular, winged, *ca* 25 cm long; raceme laxly 6-many-flowered. Floral bracts shorter than the ovary, purplish-green, linear-lanceolate, acuminate, 3-veined, *ca* 1.6 × 0.2 cm. Pedicel and ovary greenish-maroon, *ca* 2.2 cm long. Flowers drooping. Sepals and petals light green with purple veins, linear-lanceolate, acute or acuminate, 3-veined; dorsal *ca* 2 × 0.3 cm; laterals *ca* 2.2 × 0.4 cm; petals *ca* 1.8 × 0.3 cm. Lip convolute, shortly clawed, obovate, *ca* 2 × 1 cm; veins with short, fine, hyaline hairs; 3-lobed above the middle; laterals triangular; midlobe semi-elliptic or ovate, margin crenulated. Column greenish-white, broader above and rounded, elongated, *ca* 10 × 2 mm. Stigma obovate. Pollinia 2, oblong, 2 × 2 mm. Capsules drooping, spindle-shaped, *ca* 2.5 cm long.

Fl. & Fr.: June-December.

Habitat: Occasional, it is growing under shady conditions in humus soils of moist deciduous forests (600-1100 m).

Distribution: ***World*:** Australia, Bangladesh, Bhutan, Borneo, China, Fiji, India, Japan, Java, Laos, Malaysia, Myanmar, Nepal, New Caledonia, New Guinea, Niue, Ryukyu Is., Pakistan, Philippines, Samoa, Society Is., Solomon Is., Sulawesi, Sumatra, Thailand, Taiwan, Timor and Vietnam. ***India*:** Andaman & Nicobar Is., Andhra Pradesh, Arunachal Pradesh, Goa, Gujarat, Jharkhand, Karnataka, Kerala, Maharashtra, Meghalaya, Odisha, Sikkim, Tamil Nadu, Telangana, Uttarakhand and West Bengal. ***Eastern Ghats*:** *AP:* Maredumilli forest, Rampa hills, Nallamala hills and Visakhapatnam forest; *KA:* BR hills; *OD:* Brahmanigam forests, Chandrapur forest, Gopalpur-Gunupur forest, Govindapali-Gupteswara forest, Kapilas hills, Mahendragiri hills, Rebana forest and Similipal forest; *TE:* Mannanur forest; *TN:* Kolli hills.

Specimens Examined: *AP:* Betluth narava to GBM (PKM), *BSS, KP & SKB* 32331 (SKU); Maredumilly (EG), *K.N.Reddy* 76127 (FRLH); Rampa hills (EG), *MSR* 1582 (MH); Maredumilli-Kutravada (EG), *SM* 112 (RPRC); Gudem (VSKP), *BR, KP & MVS* 34681 (SKU). *OD:* Mohana (GJM), *SM* 114 (CAL); Mahendragiri hills (GJP), *SM* 182 (CAL); Jungalghar (KDM), *SM* 129 (CAL). *TN*: Kolli hills (NK), *KMM* 62249 & 62251 (RHT).

Nervilia crociformis (Zoll. & Moritzi) Seidenf., Dansk Bot. Ark. 32: 151. 1978; Misra, Orchids Orissa 244. 2004; Ramakrishna in Pullaiah, Fl. Andhra Pr. 3: 950. 1997; Misra *et al.,* Bull. Bot. Surv. India 50: 140. 2008. *Bolborchis crociformis* Zoll. & Moritzi, Syst. Verz. 89. 1846. *Nervilia prainiana* (King & Pantl.) Seidenf., Dansk Bot. Ark. 32(2): 149. 1978; Seidenfaden in K.M. Matthew, Fl. Tamilnadu Carnatic 3: 1571. 1983. *Pogonia prainiana* King & Pantl., J. Asiat. Soc. Bengal, Pt. 2, Nat. Hist. 65: 129. 1896.

Types: Java: Tjikoya, *Zollinger* 762 (W & LE).

Etymology: The specific epithet refers to the shape and colour of the flowers similar to genus *Crocus* (Iridaceae).

Terrestrial hysteranthous herbs. Corm globose, *ca* 1.5 × 1.5cm; stolons 1-2, terete, *ca* 4.5 cm long. Leaf prostrate on ground, fleshy, reniform to sub-orbicular, base deeply cordate, margins feebly undulate, apex rounded, pubescent above, glabrous with waxy texture below, *ca* 8 × 6 cm; petiole *ca* 1 cm long. Inflorescence *ca* 6 cm long, with single flower at apex. Floral bracts shorter than the ovary, pale green, ovate-lanceolate, acute, veined, *ca* 2 × 1 mm. Pedicel and ovary green, *ca* 6 mm long. Sepals and petals greenish-brown, oblong-oblanceolate, acute, 3-5-veined, *ca* 1.8 × 0.4 cm. Lip whitish-rose, convolute, sessile, obovate, *ca* 1.8 cm long, veins with violet hairs; 3-lobed near the apex; laterals oblong-ovate; midlobe sub-orbicular, margins incised crenulated, *ca* 4.5 mm broad; disc with three irregularly dentate lines. Column whitish-yellow, broader above, elongated, *ca* 1 × 0.25 cm. Stigma obovate-orbicular. Pollinia 2, clavate, 1.6 × 0.8 mm. Capsules drooping, ellipsoid, *ca* 1.4 cm long.

Fl. & Fr.: May-July.

Habitat: Rare, it is growing in dense, shady conditions in humus-rich soils of semi evergreen and moist deciduous forests (600-1100 m).

Distribution: ***World*:** Africa, Australia, Bhutan, China, India, Java, Laos, Malaysia, New Caledonia, New Guinea, Nepal, Philippines, Samoa and Sumatra. ***India*:** Andhra Pradesh, Chhattisgarh, Karnataka, Kerala, Manipur, Odisha, Sikkim, Tamil Nadu and Uttarakhand. ***Eastern Ghats:*** *AP:* Visakhapatnam forest; *OD:* Brahmanigam forests, Gopalpur-Gunupur forest, Govindapalli-Gupteswara forest, Mahendragiri hills, Rebana forest and Similipal forest; *TN:* Javvadhu hills, Kolli hills and Thiruvannamalai hills.

Specimens Examined: *AP:* Sunkarametta (VSKP), *KP* 6425 (BSID); Wongasava coffee plantation (VSKP), *SM* 113 (RPRC). *OD:* Jangalghar (KDM), *SM* 132 & 133 (CAL); Paniganda (GJP), *SM* 167 (CAL). *TN:* Peemanadhukallu (TVM), *Arumugam* 37585 (FRLH); Javvadhu hills (TVM), *RVS* 37585, *KRK* 6035 (FRLH); Kolli hills (NK), *KMM* 63066 & 63073 (RHT); Thiruvannamalai hills (TVM), *KMM* 23465 (RHT).

Nervilia falcata (King & Pantl.) Schltr., Bot. Jahrb. Syst. 45: 402. 1911; Misra Orchids Odisha 160. 2015. *Pogonia falcata* King & Pantl., J. Asiat. Soc. Bengal, Pt. 2, Nat. Hist. 65: 129. 1896.

Types: India: Duars, *Pantling* 439 (CAL! & K).

Terrestrial, hysteranthous herbs. Corm globose. Leaf green, little above the ground, broadly cordate, margins undulate, apex acuminate, glabrous, *ca* 8 cm long, petiole short. Floral bracts oblong-lanceolate. Flowers solitary, *ca* 2 cm across. Sepals and petals green with maroon flushing, linear or oblong-lanceolate, acuminate. Lip sessile, 3-lobed; laterals white, oblong; midlobe violet with white margins, obovate; ridge with glandular papillose.

Fl.: June.

Habitat: Rare, it is growing in open conditions of moist deciduous forests (ca. 800 m).

Distribution: ***World*:** Bangladesh, Bhutan and India. ***India*:** Odisha, Sikkim and Uttarakhand. ***Eastern Ghats:*** *OD:* Laxmipur forest.

Note: Reported by Misra (2015).

Nervilia gammieana (Hook.f.) Pfitzer, Nat. Pflanzenfam. 2(6): 56. 1888. *Pogonia gammieana* Hook.f., Bot. Mag. 109: t. 6671. 1883. *Nervillia gammieana* (Hook.f.) Schltr., Bot. Jahrb. Syst. 45: 404. 1911 (Illegitimate); Misra, Orchids Orissa 263. 2004.

Type: India: Darjeeling, *Gammie s.n.* (holo:K (image!)).

Etymology: This species was named in honour of James Alexander Gammie (1839-1924), who collected type material of the species.

Terrestrial, hysteranthous herbs. Corm globose. Leaf orbicular or broadly ovate, base cordate, margins undulate, *ca* 10 cm long; petiole *ca* 10 cm long. Raceme laxly 4-10-flowered. Floral bracts linear. Pedicel and ovary *ca* 8 mm long. Sepals and petals pink with whitish lines, linear-oblanceolate, *ca* 2 cm long. Lip green or yellowish-white, 3-lobed; midlobe ovate or sub-orbicular, margins crenate.

Fl.: May-June.

Distribution: ***World:*** India, Nepal and Pakistan. ***India:*** Arunachal Pradesh, Himachal Pradesh, Odisha, Sikkim, Uttarakhand and West Bengal. ***Eastern Ghats:*** *OD:* Brahmanigam and Paniganda forest.

Specimens Examined: *OD:* Brahmanigam (KDM), *SM* 162 (CAL); Paniganda (KDM), *SM* 170 (CAL).

Nervilia infundibulifolia Blatt. & McCann, J. Bombay Nat. Hist. Soc. 35: 725. 1932; Misra, Orchids Orissa 238. 2004; Misra *et al.,* Bull. Bot. Surv. India 50: 140. 2008.

Type: India: Karnataka, North Canara, Yellapur, *Bell s.n.* (BLAT).

Etymology: The specific epithet refers to the shape of the leaves (funnel-shaped).

Terrestrial hysteranthous herbs. Corm globose or sub-globose, *ca* 1 × 1 cm; stolons 2, opposite, terete, *ca* 3 cm long. Leaf little above the ground, fleshy, broadly ovate or sub-orbicular, base cordate, margins irregularly and sparingly dentate or undulate, apex acute, glabrous, green with purple bands along the veins, *ca* 3 × 4 cm; petiole *ca* 3 cm long. Inflorescence *ca* 10 cm long, with single flower at apex. Floral bracts shorter than the ovary, pale green with maroon dots, oblong-ovate, acute, *ca* 3 × 2 mm. Pedicel and ovary green with maroon dots, *ca* 5 mm long. Flowers fragrant, *ca* 2 cm across. Sepals and petals greenish-brown with maroon dots and veins, oblong-lanceolate, acuminate, 3-veined, *ca* 1.6 × 0.3 cm. Lip whitish-rose with violet patches, deflexed, shortly clawed, oblong-ovate, *ca* 1.6 × 0.3 cm, 3-lobed about the middle; laterals ovate-triangular or oblong-ovate; midlobe ovate or ovate-orbicular, margin entire or feebly denticulate; disc with median papillose, waxy white ridges. Column whitish-rose, elongated, *ca* 8 × 2.5 mm. Stigma triangular-oblong. Pollinia 2, obovate, 1.5 × 1.2 mm. Capsules drooping, ellipsoid, *ca* 1 cm long.

Fl. & Fr.: June - August.

Habitat: Rare, it grows under shady or open conditions in humus soils of moist deciduous forest (500-1100 m).

Distribution: ***World***: Bhutan, Borneo, India, Laos, Malaysia and Thailand. ***India***: Andhra Pradesh, Arunachal Pradesh, Karnataka, Kerala, Maharashtra, Odisha and Uttar Pradesh. ***Eastern Ghats:*** *AP:* Nallamala hills and Visakhapatnam forest; *OD:* Brahmanigam and Paniganda forests, Gopalpur-Gunupur forest, Mahendragiri hills and Similipal forest.

Specimens Examined: *AP:* Wongasava coffee plantation (VSKP), *SM* 0111 (RPRC). *OD:* Brahmanigam-Kirkati forest (KDM), *SM* 133 (CAL); Brahmanigam (KDM), *SM* 134 (CAL); Paniganda (GJP), *SM* 168 (CAL).

Nervilia plicata (Andrews) Schltr., Bot. Jahrb. Syst. 45: 403. 1911; C.E.C. Fisch., Fl. Madras 1458. 1928; Misra, Orchids Orissa 250. 2004; Ramakrishna in Pullaiah, Fl. Andhra Pr. 3: 950. 1997; Misra *et al.,* Bull. Bot. Surv. India 50: 140. 2008. *Arethusa plicata* Andrews, Bot. Repos. 5: 321. 1803. *Pogonia plicata* (Andrews) Lindl., Gen. Sp. Orchid. Pl. 415. 1840; Hook.f., Fl. Brit. India 6: 119. 1890. *Nervilia discolor* (Blume) Schltr., Bot. Jahrb. Syst. 45: 403. 1911.

Type: India: E. Indies, *Lambert s.n.* (holo: K).

Etymology: The specific epithet refers to the folded (plicate) leaves of the species.

Terrestrial hysteranthous herbs. Corm globose or sub-globose, *ca* 2 × 2 cm; stolons 2, terete, *ca* 8 cm long. Leaf flat on ground, plicate, orbicular or orbicular-ovate, base cordate, margins cilicate and entire, apex acuminate, greenish-maroon or brownish-green above with impressed white hairs along the veins, glabrous and purplish-green below, *ca* 8 × 8 cm; petiole *ca* 6 mm long. Inflorescence *ca* 26 cm long, two-flowered. Floral bracts shorter or longer than the ovary, pale green, lanceolate, acuminate, *ca* 1 × 0.2 cm. Pedicel and ovary greenish-maroon, *ca* 8 mm long. Sepals and petals brownish-green with brown veins, oblong-lanceolate or oblong-oblanceolate, acuminate, 3-5-veined, *ca* 2.2 × 0.5 cm. Lip whitish-purple with yellow median stripe ending before the apex, ovate or truncately rhomboid, basal margins convolute forming tube, *ca* 1.8 × 1.2 cm, obscurely 3-lobed; epichile margins crenulated. Column white, elongated, *ca*

9 × 2.2 mm, dilated apex. Stigma quadrangular or sub-orbicular. Pollinia 2, pyriform, 3 × 2 mm. Capsules drooping, ellipsoid, *ca* 2 cm long.

Fl. & Fr.: June-July.

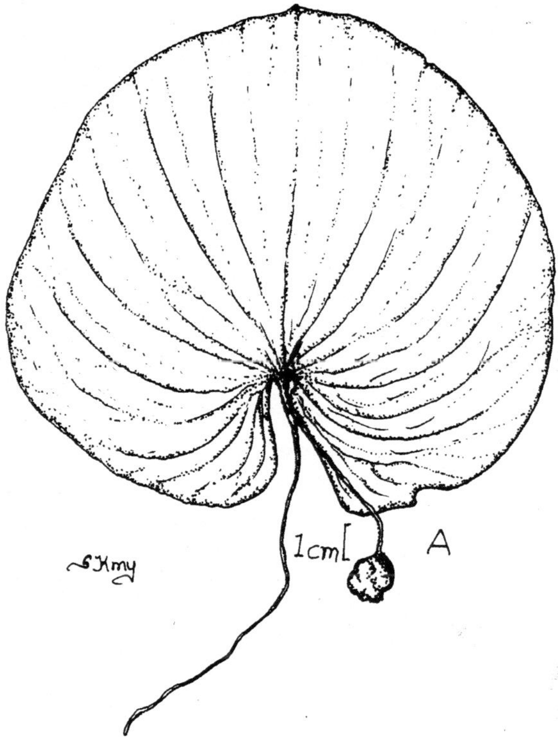

Figure 6. **Nervilia plicata** (Andr.) Schltr.
A. Habit

Habitat: It grows under deep shady conditions in humus-rich soils of moist deciduous forest (100-1200 m).

Distribution: ***World*:** Australia, Bangladesh, Bhutan, China, India, Java, Laos, Malaysia, Myanmar, Nepal, New Guinea, Pakistan, Philippines, Sumatra, Thailand, Taiwan and Vietnam. ***India:*** Andaman and Nicobar Is., Andhra Pradesh, Assam, Gujarat, Himachal Pradesh, Jammu & Kashmir, Karnataka, Kerala, Maharashtra, Manipur, Mizoram, Odisha, Sikkim, Tamil Nadu, Uttarakhand and West Bengal. ***Eastern Ghats:*** *AP:* Maredumilli forest and Tyada forest; *OD:* Paniganda forest, Chandrapur forest, Kapilas hills, Rebana forest and Similipal forest; *TN:* Javvadhu hills and Kolli hills.

Specimens Examined: *AP:* Maredumilli (EG), *VNS* 376 (CAL); Tyada RF (VSKP), *BR* & *KNR* 9891 (SKU). *OD:* Similipal forest-Lubung (MB), *SM* 1652 (CAL); Rebana forest-Lembua (KJ), *SM* 464 (CAL); Paniganda (GJP), *SM* 172 & 173 (CAL). *TN:* Javvadhu hills (TVM), *KRK* 14819 (FRLH); Thenmalai

MPCA (TVM), *KRK* 6035 (FRLH); Kolli hills, Solakkadu (NK), *KMM & VKN* 9653, *SJB* 30648 (RHT).

Doubtful species

Nervilia punctata (Blume) Makino, Bot. Mag. (Tokyo) 16: 199. 1902.

Note: Reddy *et al.* (2005) reported this species from Odisha, it has never been reported from Eastern Ghats region before and we have seen no herbarium material from this region.

Oberonia Lindl.

Gen. Sp. Orchid. Pl. 15. 1830.

World:350/India:65/EG:10

Type: *Oberonia ensiformis* (Sm.) Lindl. (*nom. cons.*).

Etymology: The genus was named after the Fairy mythical King Oberon, it refers to the strange form of the plants.

Distribution: Tropical and S. Africa to Pacific

Key to the species

1a. Scape flattened and winged or not..2

1b. Scape terete and not winged..6

2a. Lip entire..**O. proudlockii**

2b. Lip lobed..3

3a. Spike laxly flowered..**O. gammiei**

3b. Spike densely flowered..4

4a. Flowers yellowish-green or green; lip margins deeply dentate ....**O. mucronata**

4b. Flowers and lip not as above..5

5a. Petals linear; disc of lip with crescent-shaped cushion-like band.............**O. santapaui**

5b.Petals linear-lanceolate; disc not as above ....**O. brunoniana**

6a. Inflorescence shorter than the leaves..**O. ensiformis**

6b. Inflorescence longer than the leaves 7

7a. Lateral lobes of lip prominent, not auricular..**O. wightiana**

7b. Lateral lobes of lip not prominent, auricular..8

8a. Lobules of midlobe toothed or serrate..**O. verticillata**

8b. Lobules of midlobe not toothed..9

9a. Leaves oblong-lanceolate; lateral lobes of lip spreading...........................**O. falconeri**

9b. Leaves linear; lateral lobes of lip upwards..**O. pyrulifera**

Oberonia brunoniana Wight, Icon. Pl. Ind. Orient. 5: t. 1622. 1851; Hook.f., Fl. Brit. India 5: 681. 1888; C.E.C. Fisch., Fl. Madras 3: 1406. 1928; Seidenfaden in K.M. Matthew, Fl. Tamilnadu Carnatic 3: 1580. 1983; Ramakrishna in Pullaiah, Fl. Andhra Pr. 3: 952. 1997; Misra *et al.*, Bull. Bot. Surv. India 50: 140. 2008.

Syntypes: India: Tamil Nadu, Coimbatore district, Iyamally hills, *Wight s.n.* & 2941(K (images!)).

Etymology: This species is named in honour of Brunon, a British artist, who made many botanical drawings of plants for the author Nathaniel Wallich.

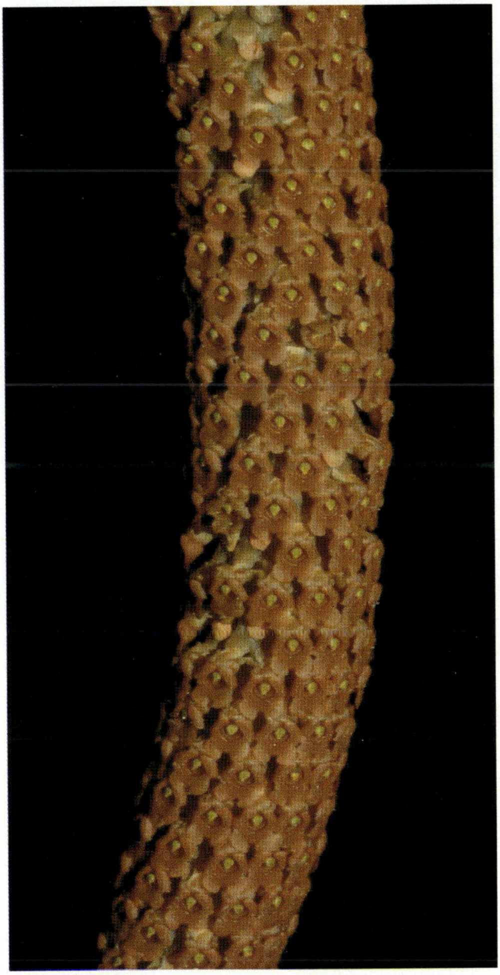

Epiphytic, erect or pendulous, acaulescent herbs. Leaves brownish-green, equitant, fleshy, oblong-ensiform, acute, *ca* 18 × 2 cm. Inflorescence adnate to the upper leaf, longer than the leaves, flattened, *ca* 32 cm long, with distinct sterile apex; spike cylindrical, densely many-flowered. Floral bracts as long as or longer than the ovary, pale brown, reflexed, ovate, irregularly denticulate, acute, gland dotted, *ca* 2 × 1.5 mm. Pedicel with ovary brownish-green, *ca* 1.5 mm long. Flowers pale brick-red or brownish-brick-red, *ca* 2.8 mm across. Sepals reflexed, entire and induplicate, obtuse or sub-acute, *ca* 2 × 1.2 mm; dorsal oblong; laterals ovate. Petals reflexed, linear-lanceolate, margins entire, apex obtuse or acute, *ca* 2 × 0.8 mm. Lip antrose, ovate, *ca* 2.5 mm long, 3-lobed; laterals ear-like, spreading, oblong or cuneate, *ca* 2.5 mm broad; midlobe 2-lobuled; lobules diverging, oblong, *ca* 1.5 mm broad, rounded or truncate; disc concave, ovate. Column yellow, sub-globose, 0.8 × 0.9 mm, beaked. Stigma rectangular. Pollinia 4, clavate, 0.4 × 0.2 mm. Capsules ellipsoid, *ca* 6 mm long.

Fl. & Fr.: September-December.

Habitat: Extremely rare, it is growing on moss covered tree trunks of semi evergreen, moist deciduous and evergreen forests (above 600 m).

Distribution: Endemic to India (Andhra Pradesh, Goa, Karnataka, Kerala, Maharashtra and Tamil Nadu). ***Eastern Ghats:*** *AP:* Rampa hills; *KA:* BR hills; *TN:* Kolli hills.

Specimens Examined: *AP:* Near Sesharayi (EG), *VNS* 613; Rampa hills (EG), *JSG* 15945 (CAL). *KA:* BR hills (CRN), *AVNR* 18210, 18211 (MH). *TN*: Kolli hills (NK), *KMM* 14623, 22822, 25373, 62652, 62742, 63302 & 4684, *KMM* & *VAA* 690, *AM* 11666, 13272 & 21676 (RHT).

Oberonia ensiformis (Sm.) Lindl., Fol. Orchid. 8: 4. 1859; Hook.f., Fl. Brit. India 5: 679. 1890; C.E.C. Fisch., Fl. Madras 1406. 1928; Misra, Orchids Orissa 333. 2004; Ramakrishna in Pullaiah, Fl. Andhra Pr. 3: 952. 1997; Misra *et al.*, Bull. Bot. Surv. India 50: 141. 2008. *Malaxis ensiformis* Sm., Cycl. 22: 14. 1812. *Oberonia iridifolia* Lindl., Gen. Sp. Orchid. Pl. 15. 1830 (Illegitimate).

Types: Nepal: Narainhetty, *Hamilton-Buchanan s.n.* (LINN & BM).

Etymology: The specific epithet refers to ensiform shape of the leaves of the species.

Tufted epiphytic herbs, pendulous, acaulescent. Leaves equitant, thickly coriaceous, fleshy, ensiform, acute or acuminate, *ca* 20 × 1.5 cm. Inflorescence shorter than the leaves, terete, *ca* 20 cm long; spike cylindrical, densely many-flowered. Floral bracts longer than the ovary, pale green, ovate, margins irregularly dentate, apex acute and deflexed, *ca* 2.5 × 1.5 mm. Pedicel with ovary pale green, *ca* 2 mm long. Flowers greenish-yellow, *ca* 1.5 mm across. Sepals reflexed, ovate, acute, *ca* 1.5 × 1.1 mm. Petals reflexed, ovate-oblong or ovate-lanceolate, margin irregularly dentate, apex obtuse, papillose, *ca* 1.5 × 1 mm. Lip antrorse, spreading, sessile, quadrate, papillose, *ca* 2 × 2 mm, 3-lobed; laterals wing like, quadrate-truncate; midlobe obcordate or obreniform, 2-lobulate; lobules rounded, converging, separated by sinus; disc with pair of median calli and terminating in pair of sub-orbicular, blister-like processes and median longitudinal groove. Column pale yellow, sub-globular, 0.4 × 0.5 mm. Stigma sub-orbicular. Pollinia 4, obovoid-oblong, 0.3 × 0.2 mm. Capsules elliptic or obovoid, *ca* 4 mm long.

Fl. & Fr.: October-May.

Habitat: Occasional, it is growing in semi-evergreen and moist deciduous forests (500-1100 m).

Distribution: ***World*:** Bhutan, China, India, Laos, Myanmar, Nepal, Thailand and Vietnam. ***India:*** Andhra Pradesh, Arunachal Pradesh, Assam, Chhattisgarh, Karnataka, Kerala, Maharashtra, Manipur, Meghalaya, Mizoram, Nagaland, Odisha, Sikkim, Tamil Nadu, Tripura, Uttarakhand and West Bengal. ***Eastern Ghats:*** *AP:* Bodhuluru, Gurtedu, Madugula forests, Gudem valley and Rampa hills; *KA:* BR hills; *OD:* Govindapalli-Gupteswara forest, Mahendragiri hills and Similipal forest; *TN*: Kolli hills.

Specimens Examined: *AP:* Rampa hills (EG), *JSG* 15945 (CAL) & *VNS* 613 (CAL); Bodhuluru, Bakkuru (EG), *SM* 103 (RPRC); Near Maddhiveedu, Tadepalli (EG), *NRR* & *DN* 84368 (MH & CAL); Gurtedu (EG), *SS* 3607; Gudem valley (VSKP), *VNS* 470, 502 (CAL); North of Gudem (Ebul reserve forest) (VSKP), *VNS* 560A (CAL); Lambasingi (VSKP), *JLE* 37179 (MH); Madugula (VSKP), *GVS* 47287 (CAL); Kotapalli (VSKP), *SM* 108 (RPRC); Sapparla (VSKP), *GVS* 29708; G.Madugula (VSKP), *GVS* 47287 (MH); Chintapalli (VSKP), *GVS* 28203 (MH); Jarada (VZN), *MV* 19835 (AU). *KA:* BR hills (CRN), *E. Barnes s.n.,* 1940 (DD). *OD:* Similipal forest-UBK (MB), *SM*

945 & 1075 (CAL); Similipal forest-Jenabila (MB), *SM* 1708 (CAL); Similipal forest-Meghasani hills (MB), *SM* 947 & 948 (CAL); Gupteswar-Ramagiri (KP), *SM* 25 & 2259 (CAL); Kutia hills (KP), *SM* 247 (CAL); Sadakna ghat (KDM), *SM* 138 (CAL). *TN:* Kolli hills (NK), *S.India Flora* 11398 (MH).

Oberonia falconeri Hook.f., Hooker's Icon. Pl. 18: t. 1780. 1888; Hook.f., Fl. Brit. India 5: 678. 1888; C.E.C. Fisch., Fl. Madras 1406. 1928; Haines, Bot. Bihar & Orissa, 1164. 1924; Misra, Orchids Orissa 343. 2004; Ramakrishna in Pullaiah, Fl. Andhra Pr. 3: 953. 1997; Misra *et al.*, Bull. Bot. Surv. India 50: 141. 2008.

Syntypes: India: Kumaon, *Falconer* 1008 (K & CAL); Dehra Dun, *Wallich* 1948 (K); Bihar, *Hooker* 120 & 630 (K); Chotanagpur, *Clarke* 21457 (K); Concan, *Law s.n.* (K).

Etymology: This species was named in honour of Hugh Falconer (1808-1865), Scottish botanist and who collected type material.

Epiphytic, erect or pendulous, acaulescent herbs. Leaves equitant, fleshy, falcate, oblong-lanceolate, acute, *ca* 8 × 1.2 cm. Inflorescence longer than leaves, ridged, terete, *ca* 15 cm long; spike cylindrical, laxly many-flowered. Floral bracts longer than the ovary, greenish-yellow, lanceolate, denticulate, acuminate, *ca* 3 × 1 mm. Pedicel with ovary yellowish-green, *ca* 2 mm long. Flowers greenish-orange or yellow, 1.3 mm across, sparsely gland dotted. Sepals reflexed, ovate, acute, 1 × 0.5 mm. Petals reflexed, oblong or linear, margins feebly undulate, acute, 1 × 0.3 mm. Lip antrorse, spreading, sub-quadrate, 1.5 × 1.5 mm, 3-lobed; laterals spreading, rounded or auricular; midlobe oblong, margin feebly waxy, 2-lobulate; lobules ovate-ligulate, acute; sinus deep, with triangular protuberance; disc cusion-like, oblong-ovate. Column sub-quadrate, 0.3 × 0.4 mm. Stigma semiorbicular. Pollinia 4, oblong-ovoid, 0.3 × 0.2 mm. Capsules ellipsoid or obovoid, *ca* 5 mm long.

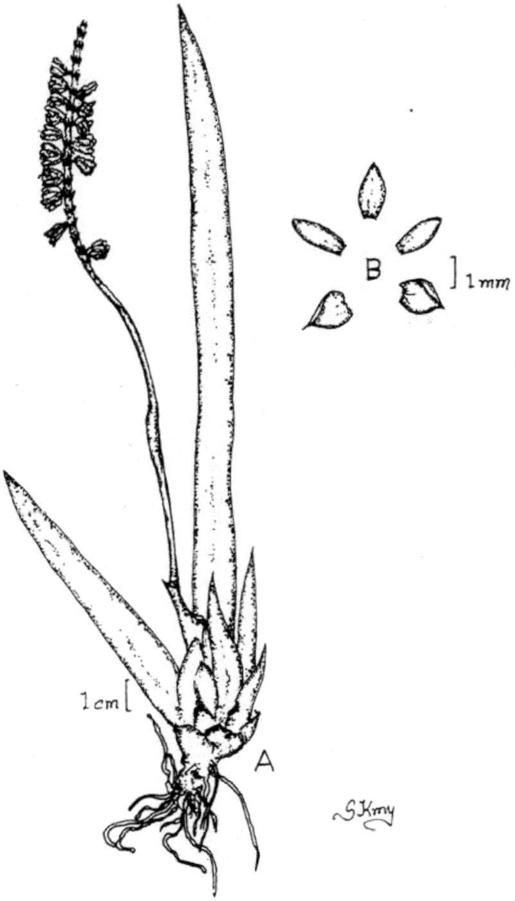

Figure 7. **Oberonia falconeri** Hook.f.
A. Habit; B. Sepals and petals

Fl. & Fr.: September-June.

Habitat: Rare, it grows on moss covered tree trunks of semi-evergreen and moist deciduous forests (above 400 m).

Distribution: ***World*:** Bhutan, China, India, Laos, Malaysia, Nepal, Thailand and Vietnam. ***India:*** Andhra Pradesh, Arunachal Pradesh, Assam, Bihar, Chhattisgarh, Gujarat, Jharkhand, Karnataka, Kerala, Madhya Pradesh, Maharashtra, Meghalaya, Odisha, Sikkim, Tamil Nadu, Tripura, Uttarakhand and West Bengal. ***Eastern Ghats:*** *AP:* Maredumilli, Sapparlagadda forest and Kotapalli forest; *OD:* Gopalpur-Gunupur forest, Rebana forest and Similipal forest.

Specimens Examined: *AP:* Way to Ukkaluru to Bodaluru, Rampa Hills (EG), *KS* 23454 (SKU); Way to Maredumilli to Karuru (EG), *KP* 6428 (BSID); Sapparlagadda (VSKP), *GVS* 42769 (MH); Kotapalli (VSKP), *SM* 104 (RPRC); Cherukonda (VSKP), *GVS* 28161 (MH); Araku (VSKP), *D. Venkata Narasaiah s.n.* (CAL). *OD:* Similipal forest-Baxua (MB), *SM* 1423 (CAL); Mohana (GJP), *SM* 120 (CAL); Gupteswar-Ramagiri (KP), *SM* 24 (CAL).

Oberonia gammiei King & Pantl., J. Asiat. Soc. Bengal, Pt. 2, Nat. Hist. 66: 578. 1897; Misra, Orchids Orissa 340. 2004.

Syntypes: India: West Bengal, Sunderbans, Shella river, *Gammie & Heinig* 92 (CAL!) & *Gammie s.n.* (K).

Etymology: This species was named in honour of James Alexander Gammie (1839-1924).

Epiphytic, erect or pendulous, acaulescent herbs. Leaves equitant, coriaceous, falcate, broadly ensiform, acuminate, *ca* 10 × 0.5 cm. Inflorescence longer than the leaves, deflexed, *ca* 30 cm long; raceme laxly many-flowered; scape compressed and narrowly winged. Floral bracts as long as or longer than the ovary, green, narrowly ovate-lanceolate, margins irregularly dentate, apex apiculate and deflexed, *ca* 3 × 1 mm. Pedicel with ovary greenish-yellow, *ca* 2 mm long. Flowers greenish-yellow, 1.5 mm across. Sepals reflexed, obtuse, 1 × 0.8 mm; dorsal ovate; laterals oblong-ovate. Petals reflexed, elliptic-oblong, margins erose, 3-veined, 0.9 × 0.4 mm. Lip spreading, deeply concave at base, broadly ovate, 3-veined, *ca* 2 × 1.4 mm, 3-lobed; laterals spreading, coarsely serrate; midlobe transversely oblong, 2-lobuled; lobules incurved, sub-quadrate, sub-entire, with sinus, sparingly punctuate; disc ovate, indistinct. Column greenish-yellow, sub-globose, 0.6 × 0.6 mm. Stigma semiorbicular. Pollinia 4, obovoid, 0.25 × 0.15 mm.

Fl.: October-November.

Habitat: Rare, it grows on moss covered tree trunks of semi evergreen and moist deciduous forests (500-1100 m).

Distribution: ***World:*** Bangladesh, China, India, Laos, Malaysia, Myanmar, Thailand and Vietnam. ***India:*** Andaman and Nicobar Is., Odisha and West Bengal. ***Eastern Ghats:*** *OD:* Similipal forest.

Specimens Examined: *OD:* Similipal forest-Barhakamada (MB), *SM* 78 (CAL); Similipal forest-UBK (MB), *SM* 981, 982, 1340 (CAL).

Oberonia mucronata (D. Don) Ormerod & Seidenf., Contr. Orchid Fl. Thailand 13: 20. 1997; Misra, Orchids Orissa 337. 2004; Misra *et al.*, Bull. Bot. Surv. India 50: 141. 2008. *O. denticulata* Wight, Icon. Pl. Ind. Orient. 5: t. 1625. 1851; Ramakrishna in Pullaiah, Fl. Andhra Pr. 3: 952. 1997. *O. iridifolia* var. *denticulata* (Wight) Hook.f., Fl. Brit. Ind. 5: 676. 1988; C.E.C. Fisch., Fl. Madras 1406. 1928. *Stelis mucronata* D. Don, Prodr. Fl. Nepal. 32. 1825.

Type: Nepal: *Buchanan-Hamilton s.n.* (holo: BM).

Etymology: The specific epithet refers to the leaves with mucronate apex.

Epiphytic, erect or pendulous, acaulescent herbs. Leaves equitant, coriaceous, fleshy, broadly ensiform or oblong-lanceolate, mucronate or acute, *ca* 10 × 2 cm. Inflorescence longer than the leaves, deflexed, *ca* 30 cm long; spike cylindrical, densely many-flowered; scape compressed, narrowly winged. Floral bracts as long as or shorter than the ovary, green, oblong-lanceolate, margins irregularly dentate, apex acute, 1.4 × 1 mm. Pedicel with ovary greenish-yellow, 1.5 mm long. Flowers yellowish-green or green. Sepals reflexed, broadly ovate, margins induplicate, apex acute, 1-3-veined, 1.2 × 0.8 mm. Petals reflexed, ovate or oblong-lanceolate, margins irregularly denticulate-crenulate, apex acute, 1-veined, 1 × 0.5 mm. Lip antrorse, sessile, quadrate or broadly ovate in outline, base concave, 1.8 × 2 mm, 3-lobed; laterals spreading, oblong, deeply dentate; midlobe transversely oblong or quadrate, 2-lobuled; lobules oblong or triangular-ovate, acute or obtuse, irregularly deeply dentate, sparingly punctuate; disc ovate, indistinct. Column greenish-yellow, sub-globose, 0.5 × 0.6 mm. Stigma semiorbicular. Pollinia 4, obovoid, 0.3 × 0.2 mm. Capsules ellipsoid, *ca* 4 mm long.

Fl. & Fr.: September-December.

Habitat: Rare, it grows in moist deciduous forests (500-1000 m).

Distribution: ***World*:** Australia, Bangladesh, Bhutan, Borneo, Cambodia, China, India, Java, Laos, Malaysia, Myanmar, Nepal, Philippines, Sulawesi, Sumatra, Thailand and Vietnam. ***India:*** Andaman & Nicobar Is., Andhra Pradesh, Arunachal Pradesh, Assam, Karnataka, Kerala, Maharashtra, Manipur, Mizoram, Meghalaya, Nagaland, Odisha, Sikkim, Tamil Nadu, Uttarakhand and West Bengal. ***Eastern Ghats:*** *AP:* Palagondi, Kotapalli forest; *KA:* BR hills; *OD:* Chandrapur forest, Kapilas hills, Rebana forest and Similipal forest.

Specimens Examined: *AP:* Palagondi (EG), *SS* 5476; Bhimavaram (VZN), *MV* 19861; Kotapalli (VSKP), *SM* 105 (RPRC); Lankapakala, R.V. Nagar (VSKP), *K.N. Reddy* 78092 (FRLH). *KA:* BR hills (CRN), *E. Barnes s.n.,* 1931, 1957 (DD). *OD:* Kendujhar forest (KJ), *HFM* 1166 (CAL); Rebana forest-Ekapadi (KJ), *SM* 705 (CAL); Similipal forest-Akhapatana (MB), *SM* 1027 (CAL).

Oberonia proudlockii King & Pantl., J. Asiat. Soc. Bengal, Pt. 2, Nat. Hist. 66: 580. 1897; C.E.C. Fisch., Fl. Madras 1406. 1928; Misra, Orchids Orissa 330. 2004; Seidenfaden in K.M. Matthew, Fl. Tamilnadu Carnatic 3: 1578. 1983.

Syntypes: India: Tamil Nadu, Nilgiri hills, Near Gudalur, *Proudlock* 602 (CAL!) & *Proudlock s.n.* (K (image!)).

Etymology: This species was named in honour of R.L. Proudlock (1892-1934), who collected type material.

Epiphytic, erect or pendulous, acaulescent herbs. Leaves equitant, not articulate, coriaceous, ovate-oblong, acute, *ca* 6 × 1 cm. Spike laxly many-flowered. Floral bracts longer than the ovary, sub-orbicular, margins lacerate. Ovary sessile or shortly pedicelled. Flowers in pits, adpressed to the rachis, greenish-yellow, 0.8 mm across. Sepals ovate; petals linear or oblong. Lip antrorse, sub-orbicular, 0.8 × 0.8 mm, margins irregularly denticulate; disc orbicular or obovate. Column globose, winged. Pollinia 4, ovoid.

Fl.: October-January.

Habitat: Rare, it grows on moss covered tree trunks of evergreen, semi ever-green and moist deciduous forests (above 1000 m).

Distribution: Endemic to peninsular India (Karnataka, Kerala, Maharashtra, Odisha and Tamil Nadu). ***Eastern Ghats:*** KA: BR hills; *OD:* Simlipal forest; *TN:* Kolli hills.

Specimens Examined: *KA:* BR hills (CRN), *E. Barnes* 2196 (DD). *OD:* Similipal forest-Meghasani hills (MB), *SM* 964, 965, 966, 967, 968, 969 (CAL). *TN:* Kolli hills (NK), *NV* & *SA* 18561, *PP* 21630 (RHT).

Oberonia pyrulifera Lindl., Fol. Orchid. 8: 3. 1859; Hook.f., Fl. Brit. Ind. 5: 678. 1988; Misra, Orchids Orissa 345. 2004.

Type: India: Khasia, Myrung, *Griffith* s.n. (K).

Etymology: The specific epithet refers to the shape of the fruits.

Epiphytic, erect, acaulescent herbs. Leaves equitant, linear. Spike longer than the leaves, laxly many-flowered. Floral bracts as long as the ovary, ovate. Pedicel with ovary 1 mm long. Flowers pale orange-yellow, 1 mm across. Sepals reflexed, ovate; petals reflexed, linear. Lip antrorse, quadrate or cuneate, 3-lobed; laterals rounded and auricular; midlobe 2-lobuled; lobules diverging, oblong or ligulate; disk indistinct.

Fl. & Fr.: September-October.

Habitat: Rare in evergreen and semi-evergreen forests (above 700 m).

Distribution: ***World:*** Bhutan, China, India and Thailand. ***India:*** Assam, Manipur, Meghalaya, Odisha and Sikkim. ***Eastern Ghats:*** *OD:* Simlipal forest.

Specimens Examined: *OD:* Similipal forest-Bandriabasa (MB), *SM* 1834 (CAL).

Oberonia santapaui Kapadia, J. Bombay Nat. Hist. Soc. 57: 265. 1960. *O. lindleyana* Wight, Icon. Pl. Ind. Orient. 5: t. 1624. 1851 (Illegitimate); Hook.f., Fl. Brit. India 5: 681. 1888; C.E.C. Fisch., Fl. Madras 1407. 1928; Seidenfaden in K.M. Matthew, Fl. Tamilnadu Carnatic 3: 1581. 1983.

Type: India: Tamil Nadu, Coimbatore, Iyamally hills, *Wight s.n.* (K).

Etymology: This species was named in honour of one of the great Indian Orchidologist, H. Santapau (1903-1970).

Epiphytic, erect or pendulous, acaulescent herbs. Leaves dirty brown, equitant, coriaceous, oblong, obtuse or acute, *ca* 15 × 2.5 cm. Inflorescence longer than the leaves, decurved, flattened, *ca* 25 cm long; spike densely many-flowered. Floral bracts longer than the ovary, reddish-green, lanceolate, acute, margins irregularly denticulate, *ca* 2.5 × 1 mm. Pedicel with ovary reddish-brown, 1.5 mm long. Flowers in verticils, reddish-brown or straw-coloured.

Sepals reflexed, oblong or oblong-ovate, obtuse, *ca* 1.3 × 0.9 mm. Petals reflexed, linear, truncate or obtuse, 1.2 × 0.3 mm. Lip antrose, ovate, shallowly crenate, *ca* 2.3 mm long, 3-lobed; laterals spreading, narrow and longitudinally elongated oblong, 2.5 mm broad; midlobe 1.5 mm broad, 2-lobuled; lobules oblong, rounded; sinus broad, with protuberance; disc ovate, with crescent-shaped cushion-like band. Column pale yellow, cylindric, 0.8 × 0.4 mm. Stigma sub-orbicular; rostellum triangular. Pollinia 4, clavate, 0.4 × 0.2 mm.

Fl.: January-March.

Habitat: Extremely rare, it grows on fully moss covered tree trunks of evergreen forests.

Distribution: Endemic to Peninsular India (Karnataka, Kerala, Maharashtra and Tamil Nadu). ***Eastern Ghats:*** *TN:* Kolli hills and Shevroy hills.

Specimens Examined: *TN:* Kolli hills (NK), *KMM* & *VAA* 857, *CR* 2007, *DIA* 7661, *KMM* & *CM* 18657 (RHT); Shevaroy hills (SLM), *NV* & *TSJ* 22445 (RHT).

Oberonia verticillata Wight, Icon. Pl. Ind. Orient. 5: t. 1626. 1851; Hook. f., Fl. Brit. India 5: 677. 1888; C.E.C. Fisch., Fl. Madras 1406. 1928; Seidenfaden in K.M. Matthew, Fl. Tamilnadu Carnatic 3: 1579. 1983.

Lectotype: India: Tamil Nadu, Nilgiri hills, Wight s.n. (BM; designated by Ansari & Balakrishnan, 1990).

Etymology: The specific epithet refers to the verticillately arranged flowers.

Epiphytic, pendulous, acaulescent herbs. Leaves equitant, coriaceous, oblong-ensiform, acute or acuminate, *ca* 15 × 1.2 cm. Inflorescence longer than the leaves, pendulous, terete, *ca* 30 cm long; spike laxly many-flowered. Floral bracts as long as or longer then the ovary, narrow-lanceolate or ovate-lanceolate, acuminate, margins irregularly dentate. Pedicel with ovary 2.5 mm long. Flowers in verticillate, orange-yellow or yellowish-green, *ca* 2 mm across. Sepals and petals reflexed, ovate or oblong-ovate, acute, 1.3 × 0.8 mm. Lip antrorse, cuneate-obovate, 2.2 mm long, 3-lobed; laterals folded upwards or spreading, rounded; midlobe 2-lobuled; lobules overlapping each other or parallel, orbicular or obovate-oblong, margins distantly serrate; sinus narrow and disc indistinct. Pollinia 4, obovoid. Capsules ellipsoid, *ca* 2.5 mm long.

Fl. & Fr.: September-October.

Habitat: Extremely rare, it grows on tree trunks and branches of evergreen forests.

Distribution: ***World:*** India and Sri Lanka. ***India:*** Karnataka, Kerala, Maharashtra and Tamil Nadu. ***Eastern Ghats:*** *TN:* Shevaroy hills.

Specimens Examined: *TN:* Shevaroy hills, Balmadies estate, Yercaud (SLM), *K. Subramanyam* 6585 (MH); Shevaroy hill top (SLM), *AVNR* 18268, 23062 (MH); Shevaroy hills (SLM), *E. Barnes s.n.* (DD); Yercaud (SLM), *DIA* 4170, 4171, 4173 & 4175, *KMM & DIA* 4960, *NV* 19106, *NV & TSJ* 22410, *KMM, SJB & NRI* 28573 (RHT).

Oberonia wightiana Lindl., Edwards's Bot. Reg. 25 (Misc.): 14. 1839; Seidenfaden in K.M. Matthew, Fl. Tamilnadu Carnatic 3: 1579. 1983. *O. arnottiana* Wight, Icon. Pl. Ind. Orient. 5: t. 1628. 1851; Ramakrishna in Pullaiah, Fl. Andhra Pr. 3: 951. 1997; Misra *et al.,* Bull. Bot. Surv. India 50: 140. 2008. *O. wightiana* sensu Hook.f., Fl. Brit. India 5: 683. 1888; C.E.C. Fisch., Fl. Madras 1407. 1928.

Lecotype: India: Tamil Nadu, Nilgiri hills, Wight 181 (K; designated by Ansari & Balakrishnan, 1990).

Etymology: This species was named in honour of R. Wight (1819-1850), well known plant collector from India and collector of type material.

Epiphytic, erect or pendulous, acaulescent herbs. Leaves equitant, coriaceous, falcate, oblong-ensiform, acute, *ca* 15 × 1 cm. Inflorescence longer than the leaves, deflexed, terete, *ca* 20 cm long; spike densely many-flowered. Floral bract longer or shorter than the ovary, ovate-lanceolate, acuminate, entire or erose, 2 × 0.8 mm. Pedicel with ovary pale green, *ca* 2 mm long. Flowers yellowish-green, *ca* 2 mm across. Sepals reflexed, ovate or ovate-oblong, obtuse or acute, 1 × 0.5 mm. Petals reflexed, linear, obtuse, 1 × 0.25 mm. Lip antrose, spreading, quadrate, 1.8 × 1.8 mm, 3-lobed; laterals wing like, oblong-orbicular or rounded, folded upwards; midlobe 2-lobuled; lobules diverging, oblong or ligulate, irregularly toothed; sinue with protuberance in between; disc ovate, indistinct. Column pale yellow, sub-globose, 0.6 mm. Stigma crescent-shaped. Pollinia 4, obovoid, 0.2 × 0.1 mm. Capsules oblong-obovoid, *ca* 3 mm long.

Fl. & Fr.: June-November.

Habitat: Rare in semi evergreen and moist deciduous forests (700-1000 m).

Distribution: ***World*:** India and Sri Lanka. ***India:*** Andhra Pradesh, Kerala and Tamil Nadu. ***Eastern Ghats:*** *AP:* Ethakonda forest and Gudem forests; *TN:* Kolli hills and Shevaroy hills.

Specimens Examined: *AP:* Ethakonda (EG), *VNS* 121 (CAL); Gudem (VSKP), *VNS* 560 (CAL). *TN:* Shevaroy hills (SLM), *AVNR* 23068 (MH); Shevaroy hills (SLM), *KMM* 62050 (RHT); Kolli hills (NK), *KMM* 62266 & 62623 (RHT).

Odisha S. Misra

Orchids India 252. 2007.

Monotypic genus

Type: *Odisha cleistantha* S. Misra

Etymology: The genus was named after the type locality state in India, Odisha.

Distribution: Endemic to India.

Odisha cleistantha S. Misra, Orchids India 252. 2007; Misra *et al.,* Bull. Bot. Surv. India 50: 141. 2008.

Type: Odisha: Mayurbhanj, Similipal forest, *SM* 2434 (holo: CAL!).

Etymology: The specific epithet refers to closed perianth of the species.

Terrestrial herbs, *ca* 40 cm high. Tubers 2, oblong or elliptic; stem slender. Leaves cauline, sessile, elliptic-lanceolate, acute, *ca* 10 × 4 cm. Spike laxly 5-10-flowered. Floral bracts longer than the ovary, oblong-lanceolate, acuminate, *ca* 2 × 4 cm. Pedicels with ovary pale green, *ca* 1.5 cm long. Flowers not opening, green. Sepals erect, ovate-lanceolate, acute, 3-veined, *ca* 1.2 × 0.4 cm. Petals erect, oblong-lanceolate, acute, *ca* 1.1 × 0.4 cm. Lip erect, distal two-third bent in, enclosed by the sepals, *ca* 1.4 mm long, 3-lobed; laterals falcate, ovate-lanceolate; midlobe oblong-lanceolate; spur laterally compressed, *ca* 1.5 cm long. Stigmatic lobes 2, oblobg; rostellum tounge-like, elongated, 3-lobed at apex; lateral lobes triangular without connection of antherophores. Auricles reduced into slender hooks. Pollinia 2, oblong, without viscidia. Capsules fusiform, *ca* 2 cm long.

Fl. & Fr.: October-December.

Habitat: Extremely rare, grows under shady condition of moist deciduous forests (350-800 m).

Distribution: Endemic to India (Andhra Pradesh, Jharkhand, Manipur and Odisha). ***Eastern Ghats:*** *AP:* Peddavalia; *OD:* Brahmanigam, Paniganda, Daringbadi and Kirikuti forests, Mahendragiri hills and Similipal forest.

Specimens Examined: *AP:* Reported by A.W. Lushington from Peddavalia (EG or VSKP). *OD:* Similipal forest (MB), *SM* 2434 (CAL); Dharmagod forest,

D.C.S. Raju 1323 (CAL); Odisha, *D.C.S. Raju s.n.* (CAL); Kendujhar (KJ), *HFM* 2824 (DD).

Pachystoma Blume

Bijdr. 376. 1825

World:20/India:2/EG:1

Type: *Pachystoma pubescens* Blume

Etymology: The generic name is derived from the Greek words, *pachy* (thick) and *stoma* (mouth), refers to the thickened mouth of the lip.

Distribution: Tropical and Subtropical Asia to SW. Pacific.

Pachystoma pubescens Blume, Bijdr. 376, t. 3, f. 29. 1825; Misra, Orchids Orissa 501. 2004. *P. senile* (Lindl.) Rchb.f., Bonplandia (Hannover) 3: 251. 1855; Hook.f., Fl. Brit. India 5: 812. 1888; C.E.C. Fisch., Fl. Madras 1426. 1928; Haines, Bot. Bihar & Orissa, 1169. 1924; Mooney, Suppl. Bot. Bihar & Orissa 210. 1956. *Apaturia senilis* Lindl., Gen. Sp. Orchid. Pl. 130. 1831.

Type: Java: Tijradjas, Krawang, *Blume s.n.* (holo: L).

Etymology: The specific epithet refers to the pubescent lip disc of the species.

Terrestrial rhizomatous herbs. Inflorescence in leafless conditions, erect, laxly few-flowered. Floral bracts longer than the ovary, linear-lanceolate or subulate-acuminate, densely puberulous outside. Pedicel and ovary ca. 6 mm long, glandular pubescent. Sepals glandular pubescent without; dorsal oblong-elliptic; laterals oblong-lanceolate. Petals narrowly oblanceolate, glabrous. Lip *ca* 1 cm long, 3-lobed; lateral lobes oblong; midlobe oblong-obovate; disk with 5 longitudinal, parallel coarse granular ridges. Column puberulous, winged. Pollinia 8, compressed, obovate.

Fl.: March-April.

Habitat: Rare, in high altitude grass lands (above 700 m).

Distribution: ***World:*** Australia, Bangladesh, Bhutan, Borneo, Cambodia, China, India, Java, Laos, Lesser Sunda Is., Malaysia, Moluccas, Myanmar, Nepal, New Caledonia, New Guinea, Philippines, Sri Lanka, Sulawesi, Sumatra, Taiwan, Thailand and Vietnam. ***India:*** Andaman Is., Arunachal Pradesh, Assam Bihar, Chhattisgarh, Jharkhand, Karnataka, Kerala, Maharashtra, Manipur, Meghalaya, Mizoram, Nagaland, Odisha, Sikkim, Tamil Nadu, Uttarakhand, Uttar Pradesh, and West Bengal. ***Eastern Ghats:*** *OD:* Gopalpur-Gunupur forest.

Specimens Exmined: *OD:* Gopalpur-Gunupur forest, *HFM* 3316 (K).

Papilionanthe Schltr.

Orchis 9: 78. 1915.

World:12/India:4/EG:2

Type: *Papilionanthe teres* (Roxb.) Schltr.

Etymology: The generic name is derived from the Greek words, *papilio* (butterfly) and *anthos* (flower), refers to the flowers like butterfly shape.

Distribution: India to SC. China and Malyasia.

Key to the species

1a. Racemes shorter than the leaves; flowers *ca* 3 cm across; midlobe not lobed.... ....**P. subulata**

1b. Racemes longer than the leaves; flowers *ca* 6 cm across; midlobe deeply bilobed..**P. teres**

Papilionanthe subulata (Willd.) Garay, Bot. Mus. Leafl. 23: 372. 1974; Seidenfaden in K.M. Matthew, Fl. Tamilnadu Carnatic 3: 1596. 1983. *Limodorum subulatum* Willd., Sp. Pl. 4: 126 1805. *Epidendrum subulatum* J. Koenig in A.J. Retzius, Observ. Bot. 6: 51. 1791 (*nom. illeg.*). *Aerides cylindrica* Lindl., Gen. Sp. Orchid. Pl.: 240. 1833 (*nom. illeg.*); Hook.f., Fl. Brit. India 6: 44. 1890; C.E.C. Fisch., Fl. Madras 1442. 1928. *Papilionanthe cylindrica* (Lindl.) Seidenf., Descr. Epidendrorum J.G. Konig 33. 1995.

Type: Sri Lanka: *Koenig s.n.* (not located).

Etymology: The specific epithet refers to the subulate shape of leaves.

Epiphytic scandent herbs. Stem erect, elongate, terete. Leaves fleshy, thick, terete or subulate, acute, *ca* 9 × 0.4 cm. Inflorescence erect, shorter than the leaves, *ca* 5 cm long, with 1-3-flowered. Floral bracts greenish-brown, triangular, acute, *ca* 2 mm long. Pedicel with ovary greenish-white with rose tinge, *ca* 3 cm long. Flowers fragrant, whitish-purple and lip with yellow spots at apex and pinkish-white spots below, *ca* 3 cm across. Sepals and petals oblong-elliptic or oblong-obovate, obtuse or rounded, 7-veined, *ca* 1.2 × 1 cm. Lip infundibuliform, 3-lobed; laterals as long as midlobe, ovate-oblong, obtuse, *ca* 6 × 5 mm; midlobe recurved, oblong, truncate; disc with fleshy ridges; spur conical, mouth funnel-shaped, *ca* 8 mm long. Column pale white, stout, winged, *ca* 4.5 × 3.4 mm; foot *ca* 4 mm long. Stigma quadrate-orbicular. Pollinia 2, globose, 1.2 × 1.2 mm. Capsules fusiform, *ca* 6 cm long.

Fl. & Fr.: March-May.

Habitat: Extremely rare, it grows in moist deciduous forests (above 600 m).

Distribution: ***World:*** India and Sri Lanka. ***India:*** Karnataka, Kerala and Tamil Nadu. ***Eastern Ghats:*** *KA*: BR hills; *TN:* Kolli hills and Shevaroy hills.

Specimens Examined: *KA:* BR hills (CRN), *AVNR* 23084 (MH). *TN:* Kolli hills, Peria shola (NK), *A. Mohan* 13246 (MH, RHT); Kolli hills (NK), *KMM & VAA* 881, *CR* 2002, *DIA* 7613, *AM* 11641, 11664 & 13246, *KMM & NV* 18336, *KMM* 22926 & 63774 (RHT); Shevaroy hills, Yercaud (SLM), *DIA* 4180, 7334 & 10282, *KMM & DIA* 4977, *KMM* 23372, 27714, 62116 & 62132 (RHT); Guthiryan (DMP), *KMM & NV* 23947 (RHT); Chitteri hills (SLM), *KMM* 62208 (RHT).

Papilionanthe teres (Roxb.) Schltr., Orchis 9: 78. 1915; Misra, Orchids Orissa 644. 2004; Ramakrishna in Pullaiah, Fl. Andhra Pr. 3: 953. 1997; Misra *et al.,* Bull. Bot. Surv. India 50: 141. 2008. *Dendrobium teres* Roxb., Fl. Ind. ed. 1832 3: 485. 1832. *Vanda teres* (Roxb.) Lindl., Gen. Sp. Orchid. Pl. 217. 1833; Hook.f., Fl. Brit. India 6: 49. 1890; C.E.C. Fisch., Fl. Madras 1444. 1928.

Type: Bangladesh: Chittagong and Silhet, *Roxburgh* 152 (BM (image!).

Etymology: The specific epithet refers to the terete shape of the leaves.

Epiphytic scandent herbs. Stem slender, terete, branched. Leaves fleshy, terete, articulate, obtuse, *ca* 15 × 0.4 cm. Inflorescence longer than leaves, extra-axillary, *ca* 30 cm long; raceme laxly 2-6-flowered. Floral bracts brown with rose dots, broadly triangular, acute, 1-veined, *ca* 3 × 1.5 mm. Pedicel with ovary whitish-rose, *ca* 4 cm long. Flowers whitish-rose, *ca* 6 cm across. Sepals and petals obovate or oblong-obovate, undulate, rounded, 9-veined; dorsal *ca* 3 × 2.3 cm; laterals *ca* 3.6 × 2.4 cm; petals *ca* 3.6 × 3.2 cm. Lip *ca* 4 × 3.8 cm, 3-lobed; laterals incurved, quadrate or quadrate-ovate; midlobe clawed, obovate-orbicular, deeply bilobed; spur conical, mouth funnel-shaped, rounded, *ca* 2 cm long. Column whitish-rose, cylindrical, with two triangular arms at apex, 1 × 0.5 cm; foot *ca* 2 mm long. Stigma orbicular-rounded. Pollinia 2, sub-globose, 3 × 2.8 mm. Capsules narrowly clavate, *ca* 10 cm long.

Fl. & Fr.: January-August.

Habitat: Rarely grows in moist deciduous forests (above 500 m).

Distribution: *World*: Bangladesh, Bhutan, Borneo, Cambodia, China, India, Java, Laos, Myanmar, Nepal, Thailand and Vietnam. ***India*:** Andaman & Nicobar Is., Andhra Pradesh, Arunachal Pradesh, Assam, Meghalaya, Mizoram, Nagaland, Odisha, Sikkim, Tamil Nadu, Tripura and West Bengal. ***Eastern Ghats:*** *AP:* Visakhapatnam hills; *OD:* Govindapalli-Gupteswara forest.

Specimens Examined: *AP:* Gudem (VSKP), *BR* & *KP* 41614 (SKU); Reported by Fischer (1928) in Visakhapatnam hills based on A.W. Lushington. *OD:* Malkanagiri, *SM* 33 (CAL).

Pecteilis Raf.

Fl. Tellur. 2: 37. 1837.

World:12/India:4/EG:2

Type: *Pecteilis susannae* (L.) Raf.

Etymology: The generic name is derived from Greek word, *pectein* (comb), refers to the pectinate lateral lobes of the lip and appears like comb.

Distribution: Russian Far East to Tropical Asia.

Key to the species

1a. Plants large, more than 1 m high; dorsal sepal broadly rhomboid................**P. gigantea**

1b. Plants small, less than 0.5 m high; dorsal sepal oblong ....**P. henryi**

Pecteilis gigantea (Sm.) Raf., Fl. Tellur. 2: 38. 1837; Misra, Orchids Orissa 140. 2004; Ramakrishna in Pullaiah, Fl. Andhra Pr. 3: 953. 1997; Misra *et al.*, Bull. Bot. Surv. India 50: 141. 2008. *Habenaria susannae* auct. non (L.) R. Br. ex Spreng. 1826; Hook.f., Fl. Brit. India 6: 137. 1890; Haines, Bot. Bihar & Orissa, 1155. 1924; Mooney, Suppl. Bot. Bihar & Orissa 206. 1956. *Platanthera susannae sensu* C.E.C. Fisch., Fl. Madras 1475. 1928, non (L.) Lind. 1835. *Orchis gigantea* Sm., Exot. Bot. 2: 79. 1806.

Types: Nepal: *Wallich* in Wall. Cat. no.7052 (P, CAL & K).

Etymology: The specific epithet refers to the giant size of the plant.

Terrestrial herbs, *ca* 1.5 m high. Tubers 1-2, ellipsoid, *ca* 8 × 3 cm; stem stout, slender, cylindrical, leafy throughout. Leaves cauline, fleshy, sessile, ovate-oblong or elliptic, acuminate, *ca* 20 × 6 cm. Inflorescence *ca* 25 cm long; raceme laxly 3-6-flowered. Floral bracts longer than the ovary, green, foliaceous, ovate-lanceolate, acuminate, *ca* 5 × 3 cm. Pedicel with ovary green, *ca* 5 cm long. Flowers fragrant, white. Dorsal sepal concave, broadly rhomboid, apex truncate, 3-veined, *ca* 6 × 3.5 cm; lateral sepals deflexed, subquadrately-oblong, acute, *ca* 5 × 2 cm. Petals falcate, linear, obtuse, 3-veined, *ca* 4.5 × 0.4 cm. Lip flabellate, *ca* 5.5 cm long, 3-lobed; laterals incurved, deeply pectinate about the middle into often bipartite; midlobe linear-spathulate, acute or obtuse; spur longer than the ovary, cylindrical, *ca* 12 cm long. Stigmatic lobes 2, spreading, parallel, oblong-subclavate; rostellum 3-lobed. Pollinia 2, oblong, *ca* 5 × 1 mm. Capsules ellipsoid, *ca* 5 cm long.

Fl. & Fr.: August-December.

Habitat: Rare, it grows under shady condition in humus rich soils of moist deciduous forest and meadow (500-1000 m).

Distribution: ***World*:** China, Java, India, Laos, Malaysia, Myanmar, Nepal, Pakistan, Thailand and Vietnam. ***India:*** Andhra Pradesh, Arunachal Pradesh, Assam, Bihar, Chhattisgarh, Himachal Pradesh, Jammu & Kashmir, Jharkhand, Karnataka, Kerala, Madhya Pradesh, Maharashtra, Manipur, Meghalaya, Mizoram, Nagaland, Odisha, Tamil Nadu, Tripura, Uttarakhand and West Bengal. ***Eastern Ghats:*** *AP:* Maredumilli forest and Madugukota forest; *OD:* Similipal forest.

Specimens Examined: *AP:* Madugukota (EG), *SS* 9912 (AU); Maredumilli to Chinturu (EG), *GVS* 68525 (MH & CAL); Maredumilli-MPCA (EG), *SM* 118 (RPRC). *OD:* Similipal forest-Patabila (MB), *SM* 1334 & 1335 (CAL).

Pecteilis henryi Schltr., Repert. Spec. Nov. Regni Veg. Beih. 4: 45. 1919; Misra, Orchids Orissa 144. 2004.

Types: China: *Henry* 12534 (K & AMES (images!)).

Etymology: This species was named in honour of A. Henry (1857-1930), who collected plants from China and also type material.

Terrestrial herbs, *ca* 40 cm high. Tubers obovoid; stem cylindric. Leaves cauline, ovate-lanceolate. Inflorescence *ca* 15 cm long, 2-flowered. Floral bracts ovate-lanceolate. Flowers cream-coloured. Dorsal sepal oblong; lateral sepals ovate-lanceolate. Petals linear. Lip broadly dome-shaped, 3-lobed; laterals pectinately fringed to about the middle and often with bipartitely filiform lobes; midlobe oblong-lanceolate; spur *ca* 9.5 cm long.

Fl.: September.

Habitat: Extremely rare, it grows under shady condition in red soils of moist deciduous forests (600-900 m).

Distribution: ***World:*** China, India, Laos, Myanmar, Thailand and Vietnam. ***India:*** Manipur and Odisha. ***Eastern Ghats:*** *OD:* Semiliguda forest.

Specimens Examined: *OD:* Semiliguda (KP), *SM* 2076 (OSC).

Pelatantheria Ridl.

J. Linn. Soc., Bot. 32: 371. 1896.

World:8/India:1/EG:1

Type: *Pelatantheria ctenoglossum* Ridl. (lectotype)

Etymology: The generic name is derived from the Greek words, *pelates* (neighbour) and *antheria* (anther), possibly refers to the stelidia that enfold the anther cap and approach each other.

Distribution: India to South Japan and West Malyasia.

Pelatantheria insectifera (Rchb.f.) Ridl., J. Linn. Soc., Bot. 32: 373. 1896; Misra, Orchids Orissa 605. 2004. *Sarcanthus insectifer* Rchb.f., Bot. Zeitung (Berlin) 15: 159. 1857; Hook.f., Fl. Brit. India 6: 68. 1890; Haines, Bot. Bihar & Orissa, 1179. 1924; Mooney, Suppl. Bot. Bihar & Orissa 215. 1956.

Type: Myanmar: Moulmein, *Parish* 267 (K).

Etymology: The specific epithet refers to the flowers shape similar to mimetic fly orchid.

Epiphytic or lithophytic, tufted herbs. Stem climbing, stout, terete, nodes with roots. Leaves coriaceous, sessile, oblong or oblong-lanceolate, base amplexicaul, apex unequally bilobulate, *ca* 6 × 1.5 cm. Racemes 2-5-flowered.

Floral bracts green, broadly triangular, obtuse, *ca* 2 × 2.2 mm. Pedicel with ovary green with purple marks, *ca* 8 mm long. Flowers *ca* 1.8 cm long. Sepals greenish-yellow with pink markings, oblong-ovate or oblong-obovate, obtuse, 3-veined; dorsal *ca* 5 × 3 mm; laterals *ca* 6.5 × 3.5 mm. Petals spreading, greenish-yellow with pink markings, oblong-oblanceolate, truncate or obtuse, 3-veined, *ca* 5 × 1.6 mm. Lip fleshy, sessile, 3-lobed; laterals white, triangular-rounded, bilobulate or obscurely lobulate; midlobe violet, porrect, ovate, truncate or rounded, *ca* 8 × 7 mm; disc with median ridge; spur white, compressed, conical, *ca* 4 mm long, mouth with hairs, back wall with callus. Column white, obovate, 2 × 1.6 mm, with long steligia. Stigma triangular-ovate and closed to column arms. Pollinia 4, sub-globose, 0.8 × 0.8 mm. Capsules clavate, *ca* 2.2 cm long.

Fl. & Fr.: October-December.

Habitat: Rare in moist deciduous forests (300-1000 m).

Distribution: ***World:*** Bangladesh, India, Myanmar, Nepal, Thailand and Vietnam. ***India:*** Assam, Bihar, Chhattisgarh, Jharkhand, Manipur, Odisha, Tripura, Uttarakhand and West Bengal. ***Eastern Ghats:*** *OD:* Brahmanigam, Paniganda, Daringbadi and Kirikuti forests, Chandrapur forest, Ghumsur forest, Gopalpur-Gunupur forest, Govindapalli-Gupteswara forest, Rebana forest and Similipal forest.

Specimens Eamined: *OD:* Bissam Katak (RG), *SM* 21 (CAL).

Peristylus Blume

Bijdr. Fl. Ned. Ind. 8: 404, t. 30. 1825 (*nom. cons.*).

World:70/India:29/EG:5

Type: *Peristylus grandis* Blume (*typ. cons.*)

Etymology: The generic name is derived from Greek words, *peri* (around) and *stylus* (column), refers to the column shape.

Distribution: Mascarenes, Tropical & Subtropical Asia to Mongolia and Pacific.

Key to the species

1a. Lip entire or obscurely 3-lobed .... **P. plantagineus**

1a. Lip perfectly 3-lobed 2

2a. Spike laxly flowered 3

2b. Spike densely flowered 4

3a. Flowers white; ovary sessile .... **P. lawii**

3b. Flowers yellowish-brown; ovary pedicelled .... **P. parishii**

4a. Lateral lobes of lip longer than the midlobe .... **P. constrictus**

4b. Lateral lobes of lip as long as or shorter than the midlobe .... **P. goodyeroides**

Peristylus constrictus (Lindl.) Lindl., Gen. Sp. Orchid. Pl. 300. 1835; Misra, Orchids Orissa 153. 2004. *Herminium constrictum* Lindl., Edwards's Bot. Reg. 18: t. 1499. 1832. *Habenaria constricta* (Lindl.) Hook.f., Fl. Brit. India 6: 161. 1890; Haines, Bot. Bihar & Orissa, 1159. 1924; Mooney, Suppl. Bot. Bihar & Orissa 207. 1956.

Types: Nepal: Wall. Cat. no. 7043 (CAL! & K).

Etymology: The specific epithet refers to the congested inflorescence.

Terrestrial herbs, *ca* 90 cm high. Tubers ovoid or ellipsoid; stem terete, slender. Leaves clustered above the summit of the stem, petioled, elliptic or elliptic-obovate, *ca* 16 cm long. Inflorescence *ca* 20 cm long; spike cylindric, densely many-flowered. Floral bracts longer than the ovary, ovate-lanceolate. Flowers horizontal, fragrant. Sepals greenish-brown; dorsal ovate-lanceolate; laterals horizontal, linear-lanceolate, margins involute. Petals white, ovate-oblong. Lip white, porrect, 3-lobed up to middle; lobes linear or linear-lanceolate; laterals longer than the midlobe; spur globose. Ovary sessile.

Fl.: June-July

Habitat: Rare, it grows under shady condition in humus rich soils of moist deciduous open forest (above 400 m).

Distribution: ***World:*** Bangladesh, Bhutan, Cambodia, China, India, Myanmar, Nepal, Philippines, Thailand and Vietnam. ***India:*** Assam, Chhattisgarh, Jammu & Kashmir, Jharkhand, Meghalaya, Punjab, Odisha, Sikkim, Uttarakhand and West Bengal. ***Eastern Ghats:*** *OD:* Chandrapur forest, Gopalpur-Gunupur forest, Rebana forest and Similipal forest.

Specimens Examined: *OD:* Similipal forest-Patabila (MB), *SM* 1329 (CAL); Similipal forest-Lulung (MB), *SM* 1795 & 1796 (CAL).

Peristylus goodyeroides (D. Don) Lindl., Gen. Sp. Orchid. Pl. 299. 1835; C.E.C. Fisch., Fl. Madras 1475. 1928; Misra, Orchids Orissa 157. 2004; Ramakrishna in Pullaiah, Fl. Andhra Pr. 3: 954. 1997; Misra *et al.,* Bull. Bot. Surv. India 50: 141. 2008; Seidenfaden in K.M. Matthew, Fl. Tamilnadu Carnatic 3: 1559. 1983. *Habenaria goodyeroides* D. Don, Prodr. Fl. Nepal. 25. 1825; Hook.f., Fl. Brit. India 6: 161. 1890; Haines, Bot. Bihar & Orissa, 1159. 1924; Mooney, Suppl. Bot. Bihar & Orissa 207. 1956.

Types: Nepal: *Wallich* in Wall. Cat. 7066A (CAL!, K, BM).

Etymology: This species was named after the orchid genus *Goodyera* (resemblance).

Terrestrial herbs. Tubers 1-2, ellipsoid or ovoid-oblong, *ca* 5 × 2 cm; stem stout, terete. Leaves 2-5, clustered slightly above the middle, sessile, elliptic-oblong or elliptic-obovate or broadly oblong, acute, *ca* 15 × 6 cm. Inflorescence *ca* 30 cm long; spike cylindrical, densely many-flowered. Floral bracts longer than ovary, green, ovate-lanceolate, acuminate, papillose, *ca* 1 × 0.3 cm. Ovary pedicelled, pale green, *ca* 6 mm long. Flowers fragrant, yellowish-green or white. Dorsal sepal concave, ovate, obtuse, 1-2-veined, *ca* 5 × 3.4 mm; lateral sepals spreading, oblong-ovate, acute, *ca* 5 × 2.5 mm. Petals broadly obovate-rounded or trapezoidal, obtuse, 1-2-veined, sparsely punctate and papillose, *ca* 4 × 4.2 mm. Lip sessile, recurved, punctuate, flabelliform, *ca* 4 × 2.6 mm, 3-lobed; laterals as long as or shorter than the midlobe, oblong; midlobe narrowly triangular; spur globular, 1.5 mm long. Stigmatic lobes 2, oblong, parallel to the auricles. Pollinia clavate or pyriform, 1.4 × 0.8 mm. Capsules ellipsoid, *ca* 1.3 cm long.

Fl. & Fr.: July-November.

Habitat: Occasional, growing under shady or open condition in humus soils of moist deciduous forests (500-1000 m).

Distribution: ***World*:** Bangladesh, Bhutan, Borneo, Cambodia, China, India, Java, Laos, Malaysia, Moluccas, Myanmar, Nepal, New Guinea, Philippines, Sri Lanka, Sulawesi, Sumatra, Taiwan, Thailand, Timor and Vietnam. ***India:*** Andhra Pradesh, Arunachal Pradesh, Assam, Bihar, Himachal Pradesh, Jharkhand, Karnataka, Kerala, Madhya Pradesh, Maharashtra, Manipur, Meghalaya, Mizoram, Nagaland, Odisha, Sikkim, Tamil Nadu, Tripura, Uttarakhand and West Bengal. ***Eastern Ghats:*** *AP:* Nallamala hills and Maredumilli forest; *OD:* Gopalpur-Gunupur forest, Mahendragiri hills, Rebana forest and Similipal forest; *TN:* Javvadhu hills and Shevaroy hills.

Specimens Examined: *AP:* Motta (KNL), *BSS, KP* & *SKB* 32352 (SKU); Gundlabrahmeswaram to Kurnool (KNL), *BR* & *KP* 34642 (SKU); Nallamalais (KNL), *JLE* 22180 (MH); Ramanapenta RF (KNL), *JLE* 32631; Maredumilli-MPCA (EG), *SM* 115 (RPRC). *OD:* Mahendragiri hills (GJM), *SM* 177 (CAL); Boragharro (GJM), *VNS* 5940 (MH); Gopalpur (KH), *SM* 421 (CAL); Dudukaguda (KH), *SM* 404 (CAL); Kaumaska (KH), *SM* 401 (CAL). *TN:* Shevaroy hills (SLM), *J. Ghatak* 724 (CAL); Shevaroy hills, Yercaud (SLM), *DIA* 10230 (RHT); Javvadhu hills (TVM), *RVS* & *KRK* 46377 (FRLH); Perriyakalrayan (SLM), *KMM* & *DIA* 5488 (RHT).

Peristylus lawii Wight, Icon. Pl. Ind. Orient. 5: t. 1695. 1851; Misra, Orchids Orissa 160. 2004. *Habenaria lawii* (Wight) Hook.f., Fl. Brit. India 6: 162. 1890; Haines, Bot. Bihar & Orissa, 1160. 1924; Mooney, Suppl. Bot. Bihar & Orissa 208. 1956.

Types: India: Karnataka, Belgaum, *Law* 68 & *s.n.* (K & CAL).

Etymology: This species was named in honour of well-known plant collector from Peninsular India, J.S. Law (1810-1885), who collected type material.

Terrestrial small herb. Tubers 1-2, ovate-oblong or ellipsoid; stem slender. Leaves clustered at the middle of the stem, sessile, oblong-elliptic or ovate-oblong, acute, *ca* 8 × 2 cm. Spike laxly many-flowered. Floral bracts longer than the ovary, lanceolate, acuminate, *ca* 1.4 × 0. 25 cm. Ovary sessile. Flowers white, *ca* 4 mm across. Dorsal sepal concave, ovate-oblong, obtuse, 1-veined, *ca* 3.2 × 1.5 mm; lateral sepals spreading, linear-lanceolate, obtuse, *ca* 3.5 × 1.5 mm.

Petals obovate, obtuse, 2-veined, *ca* 2.5 × 2 mm. Lip clawed, *ca* 3 mm long, 3-lobed about the middle; laterals shorter than the midlobe, triangular; midlobe broader than the lateral lobes, triangular; spur scrotiform. Capsules fusiform.

Fl. & Fr.: June-August.

Habitat: Rare, growing under light shady condition in humus soils of moist deciduous forests (500-800 m).

Distribution: ***World*:** India, Myanmar and Nepal. ***India:*** Andhra Pradesh, Bihar, Chhattisgarh, Gujarat, Jharkhand, Karnataka, Kerala, Madhya Pradesh, Maharashtra, Odisha, Tamil Nadu, Uttarakhand and West Bengal. ***Eastern Ghats:*** *AP:* Maredumilli forest; *OD*: Similipal forest.

Specimens Examined: *AP:* Maredumilli (EG), *K.N. Reddy & CSR* 1370 (FRLH). *OD:* Similipal forest-Lulung (MB), *SM* 1800 (CAL).

Peristylus parishii Rchb.f., Trans. Linn. Soc. London 30: 139. 1874; Misra, Orchids Orissa 164. 2004. *Habenaria parishii* (Rchb.f.) Hook.f., Fl. Brit. India 6: 161. 1890.

Type: Myanmar (Burma): Moulmein, *Parish* 216 (holo:K (image!)).

Etymology: This species was named in honour of R.C.S.P. Parish (1822-1897), who collected the type material.

Terrestrial herbs, *ca* 60 cm high. Tubers 1-2, ellipsoid or oblong-ovoid, *ca* 5 × 2.5 cm. Leaves clustered near the middle of the stem, membranous, sessile, ovate-elliptic, undulate, acute, *ca* 12 × 5 cm. Inflorescence *ca* 40 cm long; spike laxly many-flowered. Floral bracts longer than the ovary, pale brown, lanceolate, acuminate, 1-veined, *ca* 1 × 0.4 cm. Pedicel with ovary pale brown, *ca* 6 mm long. Flowers yellowish-brown, *ca* 4 mm across. Dorsal sepal concave, ovate, rounded, 1-veined, *ca* 4 × 2 mm; lateral sepals spreading, linear-oblong, margins involute, rounded, *ca* 4.5 × 2 mm. Petals ovate, rounded, 2-veined, *ca* 4 × 3 mm. Lip clawed, base concave, ovate, *ca* 5 × 3 mm, 3-lobed; laterals oblong or triangular, rounded; midlobe longer than the laterals, deflexed, triangular-ovate, rounded; spur pale yellow, broadly fusiform, *ca* 2.5 mm long. Stigmatic lobes 2, oblong or clavate. Pollinia pyriform, 2.6 × 1.5 mm. Capsules ellipsoid, *ca* 8 mm long.

Fl. & Fr.: June-August.

Habitat: Rare, it is growing under shady conditions in humus rich soils of semi evergreen and moist forests (400-1000 m).

Distribution: ***World*:** Bhutan, China, India, Myanmar, Nepal, Thailand and Vietnam. ***India:*** Andaman & Nicobar Is., Andhra Pradesh, Arunachal Pradesh, Chhattisgarh, Meghalaya, Nagaland, Odisha and Sikkim. ***Eastern Ghats:*** *AP:* Nallamala hills, Gudem; *OD:* Similipal forest.

Specimens Examined: *AP:* Nallamalais, Peddachervu (KNL), *CSR & V.S. Raju* 1456 (KU); Gudem (VSKP), *BR & KP* 41620 (SKU). *OD:* Similipal forest-Nekadanacha (MB), *SM* 1649 (CAL).

Peristylus plantagineus (Lindl.) Lindl., Gen. Sp. Orchid. Pl. 300. 1835; C.E.C. Fisch., Fl. Madras 1475. 1928; Misra, Orchids Orissa 148. 2004; Ramakrishna in Pullaiah, Fl. Andhra Pr. 3: 954. 1997; Misra *et al.,* Bull. Bot. Surv. India 50: 142. 2008. *Herminium plantagineum* Lindl., Edwards's Bot. Reg. 18: t. 1499. 1832. *Habenaria wightii* Trimen, Syst. Cat. Fl. Pl. Ceylon 91. 1885; Hook.f., Fl. Brit. India 6: 162. 1890.

Type: Sri Lanka: *Macrae s.n.* (holo: K).

Etymology: The specific epithet refers to the leaves of the species similar to European *Plantago* (Plantaginaceae).

Terrestrial herbs, *ca* 40 cm high. Tubers 1-2, oblong or oblong-ellipsoid, *ca* 4 × 2 cm; stem robust. Leaves clustered above the middle, sessile, oblong-lanceolate or elliptic-lanceolate or ovate-lanceolate, acute, *ca* 12 × 5 cm. Inflorescence *ca* 25 cm long; spike densely many-flowered. Floral bracts longer than the ovary, greenish-brown, ovate-lanceolate, acuminate, 1-veined, *ca* 6 × 2 mm. Pedicel with ovary pale brown, *ca* 6 mm long. Flowers brownish-green, *ca* 2 mm across. Sepals papillose without, rounded and mucro, 1-veined; dorsal broadly ovate, *ca* 3 × 2 mm; laterals falcate, oblong-ovate, *ca* 3 × 1.2 mm. Petals broadly ovate-suborbicular, retuse, 2-veined, *ca* 2.8 × 2.4 mm. Lip sessile, concave, broadly oblong-ovate, *ca* 2.5 × 2.2 mm, 3-veined, obscurely 3-lobed;

midlobe prominent, triangular; spur shorter than the ovary, scrotiform, 1.4 mm long. Stigmatic lobes 2, oblong. Pollinia pyriform, 0.6 × 0.4 mm. Capsules ellipsoid, *ca* 8 mm long.

Fl. & Fr.: August-October.

Habitat: Rare, it is growing under shady condition in humus rich soils of moist deciduous forest (400-1000 m).

Distribution: ***World*:** India, Nepal and Sri Lanka. ***India:*** Andhra Pradesh, Chhattisgarh, Goa, Gujarat, Karnataka, Kerala, Maharashtra, Madhya Pradesh, Odisha, Tamil Nadu and West Bengal. ***Eastern Ghats:*** *AP:* Maredumilli forest and Kotapalli forest; *OD:* Chandrapur forest, Govindapali-Gupteswara forest, Rebana forest and Similipal forest.

Specimens Examined: *AP:* Gundlabrahmeswaram (KNL), *BR* & *KP* 34672, 41622 (SKU); Way to Maredumilli to Karuru (EG), *KP* 6427 (BSID); Maredumilli (EG), *GVS* 343 (MH); Kotapalli (VSKP), *SM* 120 (RPRC). *OD:* Similipal forest-Lulung (MB), *SM* 1504 & 1677 (CAL); Similipal forest-Jenabila (MB), *SM* 844 (CAL).

Phalaenopsis Blume

Bijdr. Fl. Ned. Ind. 7: 294. 1825.

World:40/India:9/EG:1

Type: *Phalaenopsis amabilis* (L.) Blume

Etymology: The generic name is derived from Greek words, *phalaena* (Moth) and *opsis* (resemblance), refers to the flowers shape resemblance to moth.

Distribution: Tropical and Subtropical Asia to NE. Australia.

Phalaenopsis deliciosa Rchb.f., Bonplandia (Hannover) 2: 93. 1854. *Kingidium deliciosum* (Rchb.f.) H.R. Sweet, Amer. Orchid Soc. Bull. 39: 1095. 1970; Misra, Orchids Orissa 621. 2004; Ramakrishna in Pullaiah, Fl. Andhra Pr. 3: 947. 1997; Misra *et al.,* Bull. Bot. Surv. India 50: 139. 2008. *Doritis deliciosa* (Rchb.f.) T. Yukawa & K. Kita, Acta Phytotax. Geobot. 56: 156 2005. *D. wightii* (Rchb.f.) Benth. & Hook. f., Gen. Pl. 3: 574. 1883; Hook.f., Fl. Brit. India 6: 32. 1890.

Type: Java?: Tjikoya, *Zollinger* 1492 (holo:W).

Etymology: The specific epithet refers to the delicate flowers of the species.

Epiphytic herbs. Roots copious; stem stout, *ca* 2 cm long. Leaves 2-3, coriaceous, sessile, obovate-oblong or elliptic-obovate, obtuse, *ca* 20 × 5 cm. Inflorescence flexuous, *ca* 25 cm long; raceme laxly 6-12-flowered. Floral bracts pale green, triangular, acuminate, 1-veined, *ca* 6 × 2 mm. Pedicel with ovary whitish-green, *ca* 1 cm long. Flowers scented, *ca* 2 cm across. Sepals and petals white with purple markings, spreading, obtuse or rounded, 5-veined; dorsal elliptic-oblong, *ca* 10 × 3 mm; laterals obovate-oblong or oblong-ovate, *ca* 9 × 4 mm; mentum conical, *ca* 2 mm long; petals obovate, *ca* 8 × 3 mm. Lip sessile, 3-lobed; laterals white with purple markings, hatchet-shaped, *ca* 4 × 4 mm; midlobe whitish-purple or white with purple streaks, broadly triangular-ovate, base narrow, bilobed or emarginated, with 2-awned, *ca* 7 × 6.5 mm; disc with oblong callus. Column whitish-purple, oblong-ovate, *ca* 5 × 2 mm; foot 3 mm long. Stigma orbicular. Pollinia 2, globular-ovoid, 1 × 1 mm. Capsules oblong, *ca* 5 cm long.

Fl. & Fr.: June-October.

Habitat: Rare, it is growing in evergreen, semievergreen and moist deciduous forests (above 400 m).

Distribution: ***World*:** Bangladesh, Bhutan, Borneo, Cambodia, China, India, Java, Laos, Malaysia, Myanmar, Nepal, Philippines, Sri Lanka, Sulawesi, Sumatra, Thailand and Vietnam. ***India:*** Andaman and Nicobar Is., Andhra Pradesh, Arunachal Pradesh, Assam, Jharkhand, Karnataka, Kerala, Meghalaya, Odisha, Sikkim and Tamil Nadu. ***Eastern Ghats:*** *AP:* Visakhapatnam forest; *OD:* Rebana forest and Similipal forest

Specimens Examined : *OD:* Kuldiha forest (BLW), *SM* 1763 (CAL); Rebana forest-Lembua (KJ), *SM* 470 (CAL).

Pholidota Lindl. *ex* Hook.

Exot. Fl. 2: t. 138. 1825.

World:29/India:12/EG:2

Type: *Pholidota imbricata* Lindl.

Etymology: The generic name is derived from the Greek word, *pholidos* (scale), refers to the scaly sheathing floral bracts.

Distribution: Tropical and Subtropical Asia to SW. Pacific.

Key to the species

1a. Leaf thin; lateral sepals connate at base...... ....**P. pallida**

1b. Leaf thick; lateral sepals free..**P. imbricata**

Pholidota imbricata Lindl., Exot. Fl. 2: t. 138. 1825; Hook.f., Fl. Brit. India 5: 845. 1888; C.E.C. Fisch., Fl. Madras 3: 1431. 1928; Mooney, Suppl. Bot. Bihar & Orissa 210. 1956; Misra, Orchids Orissa 512. 2004; Misra *et al.*, Bull. Bot. Surv. India 50: 142. 2008.

Lectotype: Nepal: Hooker icon. 138 (Exot. Fl. 1825; designated by Seidenfaden, 1986).

Etymology: The specific epithet refers to the flower arrangement on rachis of the species.

Epiphytic herbs. Rhizomes creeping, woody; pseudobulbs aggregated, green, narrowly oblong-ovate, *ca* 5 × 3 cm. Leaf thick, sub-plicate, oblong-elliptic or broadly obovate, acute, *ca* 26 × 8 cm; petiole grooved, *ca* 3 cm long. Inflorescence arising from the base of petiole, longer than the leaves, drooping, *ca* 40 cm long; raceme densely many-flowered. Floral bracts longer than the ovary, pale brown, convolute, broadly ovate, acute, 12-veined, gland dotted, *ca* 1.2 × 1 cm. Pedicel with ovary brown, *ca* 6 mm long. Flowers bifarious, white, *ca* 8 mm across. Dorsal sepal concave, elliptic-ovate, rounded, 5-veined, *ca* 6 × 5 mm; lateral sepals concave, sickle-shaped, acute, keeled, *ca* 7 × 4 mm. Petals linear, acute, 1-veined, *ca* 6 × 2 mm. Lip sessile, base concave, panduriform, 9-veined; hypochile yellow, with three longitudinal ridges within; laterals hatchet-shaped, *ca* 2.5 × 2.5 mm; midlobe deflexed, 2-lobulate; lobules auricular, rounded, with broad sinus in between, *ca* 2.5 × 1.4 mm. Column whitish-brown, oblong, winged, *ca* 3 × 2.5 mm. Stigma funnel-shaped. Pollinia 4, ovoid-pyriform, 0.9 × 0.8 mm. Capsules ellipsoid, *ca* 2 cm long.

Fl. & Fr.: November-March.

Habitat: Rare, it is growing in semi evergreen/moist deciduous forests (600-1400 m).

Distribution: ***World*:** Australia, Bangladesh, Bhutan, Borneo, Cambodia, China, Fiji, India, Java, Laos, Malaysia, Moluccas, Myanmar, Nepal, New Caledonia, New Guinea, Philippines, Singapore, Solomon Island, Sri Lanka, Sulawesi, Sumatra, Thailand, Tibet and Vietnam. ***India:*** Andaman & Nicobar Is., Andhra Pradesh, Arunachal Pradesh, Assam, Bihar, Chhattisgarh, Jharkhand, Karnataka, Kerala, Maharashtra, Manipur, Meghalaya, Mizoram, Nagaland, Odisha, Sikkim, Tamil Nadu, Tripura, Uttarakhand and West Bengal. ***Eastern Ghats:*** *AP:* Kotapalli forest, Bison hills; *OD:* Govindapali-Gupteswara forest, Kapilas hills, Mahendragiri hills, Rebana forest and Similipal forest; *TN:* Kolli hills.

Specimens Examined: *AP:* Bison hill (EG), *CAB* 5050 (MH); Kotapalli (VSKP), *SM* 120 (RPRC). *OD:* Similipal forest (MB), *Bahali & Agrawala* 430 & 623 (CAL). *TN*: Kolli hills (NK), *AM* 13275, *PP* 21751, *KMM* 22841 (RHT).

Pholidota pallida Lindl., Edwards's Bot. Reg. 21: t. 1777. 1835; Ramakrishna in Pullaiah, Fl. Andhra Pr. 3: 955. 1997; Misra *et al.*, Bull. Bot. Surv. India 50: 142. 2008; Seidenfaden in K.M. Matthew, Fl. Tamilnadu Carnatic 3: 1574. 1983.

Neotype: Nepal: Gossainthan, *Wallich s.n.* (K; designated by Halle, 1977).

Etymology: The specific epithet refers to the pale colour flowers of the species.

Epiphytic herbs; pseudobulb narrowly ovate-oblong. Leaf thin, petioled, elliptic-lanceolate or obovate-oblong, *ca* 20 × 4.5 cm. Racemes drooping, densely many-flowered. Floral bracts longer than the ovary, broadly ovate-rounded, 20-veined. Pedicel with ovary brown, *ca* 5 mm long. Flowers bifarious, creamish-white. Dorsal sepal elliptic-ovate, 5-veined; lateral sepals connate at base, ovate-oblong. Petals oblong-ovate. Lip panduriform, 3-lobed; hypochile yellow, with three longitudinal ridges within; laterals hatchet-shaped or broadly oblong-rounded; midlobe 2-lobulate; lobules quadrate-suborbicular. Pollinia 4, ovoid-pyriform. Capsules ellipsoid, *ca* 1.5 cm long.

Fl. & Fr.: June-October.

Habitat: Rare, it is growing on tree trunks of moist deciduous forest (about 600 m).

Distribution: ***World*:** Australia, Bhutan, Borneo, Laos, Malaysia, Myanmar, Nepal, Thailand and Vietnam. ***India:*** Andhra Pradesh, Arunachal Pradesh, Assam, Meghalaya and Sikkim. ***Eastern Ghats:*** *AP:* Bison hill and Pedda konda.

Specimens Examined: *AP:* Bison hill (EG), *CAB* 50376 (MH); Pedda konda (EG), *SS* 3887 (AU).

Note: In Madras herbarium (MH), one of the herbarium sheet labeled as a *Pholidota articulata* Lindl. (Gilbe's estate, Yercaud (SLM), *AVNR* 26917) but we have a doubt regarding its identity.

Phreatia Lindl.

Gen. Sp. Orchid. Pl.: 63. 1830.

World:200/India:3/EG:1

Type: *Phreatia elegans* Lindl.

Etymology: The generic name is derived from the Greek word, *phreatia* (a well) in reference to the structure formed by the lateral sepals and lip

Distribution: Tropical & Subtropical Asia to Pacific.

Phreatia elegans Lindl., Gen. Sp. Orchid. Pl. 63 1830; Hook.f., Fl. Brit. India 5: 810. 1888; Panda & Misra, J. Econ. Taxon. Bot. 37: 478. 2013.

Types: Sri Lanka (Ceylon): Peradeniya, *Macrae* 38 (K & BM (images!)).

Etymology: The specific epithet refers to the elegant (elegans) nature.

Epiphytic herbs, with very short, densely tufted, subpendulous stems. Leaves coriaceous, sessile, linear, obtuse, narrowed at base, *ca* 10 cm long. Inflorescence terminal, longer than the leaves, *ca* 15 cm long; raceme densely many-flowered. Floral bracts ovate-oblong, acuminate, 1-veined. Pedicel with ovary 2 mm long. Flowers minute, white. Sepals broadly triangular, acute, 1-veined, *ca* 2.5 mm long; lateral sepals gibbous at the saccate base and adnate to the foot of the column. Petals shorter than the sepals, broadly ovate, *ca* 1 mm long. Lip small, clawed, nearly orbicular, pilose within, *ca* 1 mm long. Column short, truncate, with a 2-horned at apex.

Fl.*:* August-September.

Habitat: Rare, it grows in high altitude shola forests (above 1000 m)

Distribution: ***World*:** Bhutan, India, Nepal and Sri Lanka. ***India:*** Kerala, Meghalaya, Sikkim, Tamil Nadu and West Bengal. ***Eastern Ghats: TN:*** Kolli hills

Specimens Examined: *TN:* Kolli hills (NK), *P.C. Panda & SM s.n.* (OSC).

Pinalia Lindl.

Orchid. Scelet. 14, 21. 1826.

World:160/India:18/EG:1

Type: *Pinalia alba* Buch.-Ham. *ex* Lindl.

Etymology: The genus was named in honour of French botanist Chevalier Pinal (1873-1956).

Distribution: Tropical and Subtropical Asia to SW. Pacific.

Pinalia polystachya (A. Rich.) Kuntze, Revis. Gen. Pl. 2: 679. 1891. *Eria polystachya* A. Rich., Ann. Sci. Nat., Bot. II, 15: 20. 1841; Hook.f., Fl. Brit. India 5: 792. 1890; C.E.C. Fisch., Fl. Madras 1425. 1928; Seidenfaden in K.M. Matthew, Fl. Tamilnadu Carnatic 3: 1577. 1983.

Type: India: Tamil Nadu, Nilgiri hills, Neddoubetta, *Perrottet s.n.* (P).

Etymology: The specific epithet refering to the many inflorescences for plant of the species.

Epiphytic herbs. Pseudobulbs caespitose, sub-cylindrical, oblong, with many nodes, *ca* 10 cm long. Leaves 3-5, membranous, sessile, elliptic-oblong or elliptic-lanceolate or linear-oblong, acute, *ca* 12 × 2.5 cm. Rachis, pedicel with ovary and sepals outside densely pubescent. Inflorescence shorter than the leaves, sub-erect, *ca* 8 cm long; raceme laxly many-flowered. Floral bracts longer than the ovary, pale yellow, deflexed, ovate-lanceolate, acuminate, glabrous, 1-veined, *ca* 8 × 2.5 cm. Pedicel with ovary pale yellow, *ca* 5 mm long. Flowers pale yellow or pale white, *ca* 6 mm long. Sepals oblong-lanceolate, acute or obtuse, 3-veined, *ca* 7 × 2.2 mm; mentum short. Petals falcate, oblong-linear or oblong-lanceolate, obtuse, glabrous, 3-veined, *ca* 6 × 1 mm. Lip sessile or shortly clawed, oblong-lanceolate or ovate sub-panduriform, obtuse, 3-veined, *ca* 5 × 2.2

mm, obscurely 3-lobed. Column yellowish-white, widening towards apex, winged, with 2 triangular stelidia, *ca* 4 × 2.5 mm; foot *ca* 3 mm long. Stigma quadrate-suborbicular. Pollinia 8, pyriform, 1 × 0.6 mm.

Fl.: October-November.

Habitat: Rare, it is growing on fully moss covered tree trunks of evergreen forest (1000-1400 m).

Distribution: Endemic to Kerala and Tamil Nadu. ***Eastern Ghats:*** *TN:* Shevaroy hills.

Specimens Examined: *TN:* Shevaroy hills, Yercad (SLM), *C. Paramasivan* 4281 (RHT).

Polystachya Hook.

Exot. Fl. 2: t. 103. 1824 (*nom. cons.*).

World:200/India:1/EG:1

Type: *Polystachya luteola* Hook.

Etymology: The generic name is derived from Greek words, *polys* (many) and *stachys* (spike), refers to nature of inflorescence.

Distribution: Tropical and Subtropical.

Polystachya concreta (Jacq.) Garay & H.R. Sweet, Orquideologia 9: 206. 1974; Misra, Orchids Orissa 564. 2004; Ramakrishna in Pullaiah, Fl. Andhra Pr. 3: 955. 1997; Misra *et al.,* Bull. Bot. Surv. India 50: 142. 2008. *Epidendrum concretum* Jacq., Enum. Syst. Pl. 30. 1760. *Polystachya purpurea* Wight, Icon. Pl. Ind. Orient. 5: t. 1679. 1851; Hook.f., Fl. Brit. India 6: 21. 1890; C.E.C. Fisch., Fl. Madras 1437. 1928.

Neotype: Martinique: *Privault* 136 (P; designated by Mytnik-Ejsmont & Baranow, 2010).

Epiphytic herbs. Pseudobulbs clustered, green, compressed, narrowly ovate, *ca* 3 × 1 cm; stem short, compressed. Leaves 2-5, coriaceous, plicate, sessile, linear-oblong or oblong-elliptic to lanceolate, acute, *ca* 12 × 2.5 cm. Inflorescence *ca* 20 cm long, hairy; panicles laxly many-flowered. Floral bracts pale green, triangular, acuminate, 3-veined, *ca* 2 × 2 mm. Pedicel with ovary greenish-yellow, ca 6 mm long. Flowers non-resupinate, greenish-yellow, *ca* 4 mm across. Dorsal sepal elliptic-oblong, acute, 4-veined, *ca* 4 × 2 mm; lateral sepals spreading, triangular-ovate, acute, *ca* 4 × 3 mm; mentum conical, 2 mm long. Petals linear or linear-oblanceolate, obtuse, 1-veined, *ca* 4 × 0.8 mm. Lip sessile, cuneate, broadly oblong-obovate in outline, *ca* 5 × 3 mm, 3-lobed; laterals ovate-triangular; midlobe recurved, broadly ovate, cuspidate or rounded, crenulate; disc yellow, obovate, farinose. Column greenish-yellow, oblong, *ca* 3 × 1.2 mm; foot *ca* 3 mm long. Stigma broadly sub-orbicular. Pollinia 4, sub-globose, 0.9 × 1 mm. Capsules elliptic-oblong, *ca* 2 cm long.

Fl. & Fr.: February-April.

Habitat: Rare, it is growing on moss covered tree trunks of evergreen and semi-evergreen forests (800-1400 m).

Distribution: ***World*:** Argentina, Bolivia, Borneo, Brazil, China, Colombia, Cuba, Ecuador, Florida, Guatemala, Honduras, India, Java, Laos, Malaysia, Mexico, Moluccas, Panama, Peru, Philippines, Singapore, Sri Lanka, Sulawesi, Sumatra, Thailand, Timor, Venezuela and Vietnam. ***India:*** Andaman & Nicobar Is., Andhra Pradesh, Arunachal Pradesh, Karnataka, Kerala, Maharashtra, Odisha and Tamil Nadu. ***Eastern Ghats:*** *AP:* Cherukonda hills; *KA:* BR hills; *OD*: Mahendragiri hills; *TN:* Kolli hills and Shevaroys hills.

Specimens Examined: *AP:* Galikonda (VSKP), *BR* & *KP* 34649 (SKU); Sambarikonda (VSKP), *BR, KP* & *MVS* 34692 (SKU); Cherukonda (VSKP), *GVS* 29728 (MH); On way to Ontali (VSKP), *GVS* 42647 (MH). *KA:* BR hills (CRN), *MVS* & *V.S. Rao* 32477 (SKU), *AVNR* 18212 (MH). *OD:* Mahendragiri hills (GJP), *SM* 147 (CAL), *VNS* 5742 (MH). *TN:* Guthirayanbetta, Kodagaari (DMP), *TRS* 95509 (MH); Shevaroy hills, Sanyasimalai (SLM), *AVNR* 23060 (MH): Kolli hills, Way to Semmadu (NK), *N. Ahmed Sherif* & Party *s.n.* (MH); Kolli hills (NK), *MBV* 3226 & 17210 (FRLH), *KMM* & *VAA* 768, 660, 689 & 3105, *KMM* 4725, *CR* 2063, *DIA* 7562 & 7607, *AM* 11637, 13210 & 13258,

KMM & *CM* 18258 (RHT); Yercaud (SLM), *KMM* & *DIA* 2789, *DIA* 10280; Periyakalyrayan (SLM), *KMM* 15260 (RHT).

Pomatocalpa Breda

Gen. Sp. Orchid. Asclep. 3: t. 15. 1827.

World:13/India:7/EG:2

Type: *Pomatocalpa spicatum* Breda

Etymology: The generic name is derived from Greek words, *poma* (cup) and *calpa* (jug shape), refers to the shape of the lip at base.

Distribution: Tropical and Subtropical Asia to SW. Pacific.

Key to the species

1a. Ovary sessile; disc of lip without ridges.. **P. decipiens**

1b. Ovary pedicelled; disc of lip with two ridges..**P. spicatum**

Pomatocalpa decipiens (Lindl.) J.J. Sm., Natuurw. Tijdschr. Ned.-Indie 72: 103. 1912; Misra, Orchids Orissa 601. 2004. *Cleisostoma decipiens* Lindl., Edwards's Bot. Reg. 30: 11. 1844; Hook.f., Fl. Brit. India 6: 75. 1890.

Type: Sri Lanka (Ceylon): *Fielding s.n.* (holo:K (image!)).

Etymology: The specific epithet refers to this species easily mistaken by *Smitinandia micrantha* (*Saccolabium micrantha*).

Epiphytic herbs. Leaves coriaceous, lorate. Racemes shorter than the leaves, simple or branched, densely many-flowered. Floral bracts broadly triangular. Flowers yellow with or without reddish blotches, fleshy, *ca* 7 mm long. Sepals oblong-obovate or ovate. Petals elliptic-ovate or oblong-obovate. Lip 3-lobed; laterals broadly truncate; midlobe sub-orbicular or broad ovate; spur prominent jug-shaped, rounded. Ovary sessile. Capsules ellipsoid.

Fl. & Fr.: January-May.

Habitat: Rare, it is growing in moist deciduous forests (400-600 m).

Distribution: ***World:*** Bangladesh, India and Sri Lanka. ***India:*** Odisha. ***Eastern Ghats:*** *OD:* Barbara-Dhuanali forest and Kapilas hills.

Specimens Examined: OD: Kapilas hills- Barbara-Dhuanali forest, *SM* 800 & 2141 (OSC).

Pomatocalpa spicatum Breda, Kuhl & Hasselt, Gen. Sp. Orchid. Asclep. t. 15. 1827; Misra *et al.,* Bull. Bot. Surv. India 50: 142. 2008. *P. wendlandorum* (Rchb. f.) J.J. Sm., Natuurw. Tijdschr. Ned.-Indie 72: 108. 1912; Ramakrishna in Pullaiah, Fl. Andhra Pr. 3: 956. 1997. *P. mannii* (Rchb.f.)

J.J. Sm., Natuurw. Tijdschr. Ned.-Indie 72: 105. 1912; Ramakrishna in Pullaiah, Fl. Andhra Pr. 3: 956. 1997; Misra *et al.*, Bull. Bot. Surv. India 50: 142. 2008. *Cleisostoma wendlandorum* Rchb.f., Allg. Gartenzeitung 24: 219. 1856; Hook.f., Fl. Brit. India 6: 74. 1890. *C. mannii* Rchb.f., Flora 55: 274. 1872; Hook.f., Fl. Brit. India 6: 74. 1890; C.E.C. Fisch., Fl. Madras 1448. 1928.

Lectotype: Breda, Kuhl & Hasselt (Gen. Sp. Orchid. Asclep.) t. 15. 1829; designated by Pearce & Cribb, 2002.

Etymology: The specific epithet refers to the nature inflorescence.

Epiphytic herbs; roots copious; stem short. Leaves coriaceous, falcate, sessile, linear, unequally bilobed, *ca* 14 × 3 cm. Inflorescence shorter than the leaves, simple or branched, pendulous, *ca* 8 cm long; spike densely many-flowered. Floral bracts brown, triangular, acuminate, 1-veined, *ca* 1 × 1.2 mm. Pedicel with ovary yellowish-green, *ca* 2.5 mm long. Flowers yellowish-white with or without reddish markings, *ca* 5 mm across. Sepals incurved at apex, obovate-oblong or obovate-elliptic, obtuse, 3-veined, *ca* 5 × 2 mm. Petals obovate, acute, 3-veined, *ca* 4 × 2 mm. Lip fleshy, *ca* 4.5 mm long, 3-lobed; laterals small or obscure, triangular; midlobe spreading, orbicular, rounded, entire or feebly serrulate, 2 × 1.8 mm; disc with two ridges; spur jug-shaped, sub-globose, with callus on the back wall, *ca* 1.5 mm long. Column yellow, oblong, *ca* 1.2 × 1 mm. Stigma sub-quadrate. Pollinia 4, globose, 0.8 × 0.8 mm. Capsules linear-oblong, *ca* 2 cm long.

Fl. & Fr.: March-August.

Habitat: Rare, it is growing on smooth tree trunks and branches in moist deciduous forest (600-1000 m).

Distribution: ***World***: Bhutan, Borneo, China, India, Java, Laos, Malaysia, Myanmar, New Guinea, Philippines, Sri Lanka, Sulawesi, Sumatra, Thailand and Vietnam. ***India:*** Andaman and Nicobar Is., Andhra Pradesh, Arunachal Pradesh, Assam and Sikkim. ***Eastern Ghats:*** *AP:* Pachenda hills and Palakonda hills.

Specimens Examined: *AP:* Pachenda hills, Jumlakota (VSKP), *JSG* 13947 (CAL); Palakonda hills (VSKP), *VNS* 121 (MH).

Porpax Lindl.

Edwards's Bot. Reg. 31(Misc.): 62. 1845.

World:14/India:7/EG:1

Type: *Porpax reticulata* Lindl.

Etymology: The generic name is derived from Greek word, *Porpax* (handle of shield), refers to Lip like a handle of a shield.

Distribution: India to SC. China and Borneo.

Porpax reticulata Lindl., Edwards's Bot. Reg. 31: 62. 1845; C.E.C. Fisch., Fl. Madras 1422. 1928. *Eria reticulata* (Lindl.) Benth. in Benth. & Hook.f., Gen. Pl. 3: 509. 1883; Hook.f., Fl. Brit. India 5: 786. 1890. *Aggeranthus marchantioides* Wight, Ic. t. 1737. 1851. *Porpas papillosa* Blatt. & McCann, J. Bombay Nat. Hist. Soc. 35: 268. 1931.

Type: India: East Indies, *s.coll. s.n.* (not located).

Etymology: The specific epithet refers to reticulate veined sheath of the species.

Epiphytic or lithophytic herbs; pseudobulbs aggregate, covered with reticulate veined sheaths, discoid, *ca* 3 × 1.5 cm. Leaves caducous, broadly oblong or spathulate, obtuse, *ca* 3 × 0.6 cm. Flowers solitary from the base of pseudobulbs or between the leaves, purple-brown or brownish-red. Floral bracts sub-orbicular, retuse, denticulate. Sepals united and form sub-campanulate tube, apex 3-lobed, densely papillose inside, *ca* 1.5 cm long; lobes ovate, acute. Petals included, spathulate, acute, papillose, *ca* 8 × 3 mm. Lip articulated, shortly clawed, panduriform, *ca* 6 × 4 mm, with basal tooth-like process, crenulated, 3-lobed; laterals broadly ovate; midlobe sub-orbicular, papillose. Column wingless and with long foot. Pollinia 8.

Fl.: April-June.

Habitat: Extremely rare, it grows on fully moss covered tree trunks or rocks of evergreen forest (above 1000 m).

Distribution: ***World:*** India, Laos, Thailand and Vietnam. ***India:*** Karnataka, Kerala, Maharashtra, Tamil Nadu. ***Eastern Ghats:*** *TN:* Kolli hills.

Rhynchostylis Blume

Bijdr. 7: 285, t. 49. 1825.

World:4/India:2/EG:1

Type: *Rhynchostylis retusa* (L.) Blume (lectotype)

Etymology: The generic name is derived from Greek words, *rhynchos* (beak) and *stylos* (column), referring to the beaked column.

Distribution: India to SC. China and Malaysia.

Rhynchostylis retusa (L.) Blume, Bijdr. 286. 1825; Hook.f., Fl. Brit. India 6: 32. 1890; C.E.C. Fisch., Fl. Madras 1440. 1928; Haines, Bot. Bihar & Orissa, 1178. 1924; Mooney, Suppl. Bot. Bihar & Orissa 214. 1956; Misra, Orchids Orissa 640. 2004; Seidenfaden in K.M. Matthew, Fl. Tamilnadu Carnatic 3: 1597. 1983; Ramakrishna in Pullaiah, Fl. Andhra Pr. 3: 957. 1997; Misra *et al.*, Bull. Bot. Surv. India 50: 142. 2008. *Epidendrum retusum* L., Sp. Pl. 953. 1753.

Lectotype: India: Malabar, Rheede, Hort. Malab. (ansejeli-maravara) 12: t. 1.1703; designated by Majumdar & Bakshi, 1979.

Etymology: The specific epithet refers to the leaves and lip apex nature.

Epiphytic herbs; stem sub-erect or pendulous, stout. Leaves coriaceous, sessile, linear, unequally bilobed, *ca* 25 × 2.5 cm. Inflorescence longer than the leaves, drooping, *ca* 40 cm long; raceme cylindrical, densely many-flowered. Floral bracts greenish-pink, triangular, 3-veined, *ca* 4 mm long. Pedicel with ovary whitish-rose, *ca* 2 cm long. Flowers fragrant, white with violet blotched, *ca* 2 cm across. Dorsal sepal oblong-ovate or elliptic, obtuse, 3-5-veined, *ca* 9 × 5 mm; lateral sepals broadly ovate, obtuse, *ca* 10 × 8 mm. Petals falcate, elliptic, obtuse, 3-veined, *ca* 10 × 4 mm. Lip *ca* 1.2 cm long; hypochile clawed and deflexed, deeply saccate; epichile curved upward and forward, narrowly obovate or spathulate, abruptly acute or slightly retuse, 5-veined; spur laterally compressed, oblong, *ca* 8 mm long. Column yellow, oblong, *ca* 2 × 2 mm. Stigma quadrate-orbicular. Pollinia 2, sub-globose, 1 × 1 mm. Capsules obovoid or ellipsoid, *ca* 2 cm long.

Fl. & Fr.: March-July.

Habitat: Occasional, it is growing on tree trunks and branches in moist deciduous forest (above 400 m).

Distribution: ***World:*** Bangladesh, Bhutan, Borneo, Cambodia, China, India, Java, Laos, Malaysia, Myanmar, Nepal, Philippines, Sri Lanka, Sumatra, Thailand and Vietnam. ***India:*** Andaman & Nicobar Is., Andhra Pradesh, Arunachal Pradesh, Assam, Bihar, Chhattisgarh, Himachal Pradesh, Jharkhand, Jammu & Kashmir, Karnataka, Kerala, Maharashtra, Meghalaya, Mizoram, Nagaland, Odisha, Punjab, Sikkim, Tamil Nadu, Uttarakhand and West Bengal. ***Eastern Ghats:*** *AP:* Maredumilli forest and Visakhapatnam forest; *KA:* BR hills; *OD:* Occasionally found in Eastern Ghats of Odisha; *TN:* Shevaroy hills.

Specimens Examined: *AP:* Maredumilli (EG), *BR* & *KP* 34660 (SKU), *GVS* 24251 (MH); Sutalavada (EG), *SS* 8457; Rinthada (VSKP), *GVS* 30110; Kotapalli (VSKP), *SM* 88 (RPRC). *OD:* Similipal forest-Podadiha (MB), *Bahali* & *Agrawala* 434 (CAL); Mahendragiri hills (GJP), *SM* 150 (CAL). *TN:* Shevaroy hills (SLM), *AVNR* 23071 (MH).

Satyrium Sw.

Kongl. Vetensk. Acad. Nya Handl. 21: 214. 1800 (*nom. cons.*).

World:100/India:18/EG:1

Type: *Satyrium bicorne* (L.) Thunb.

Etymology: The generic name is derived from Greek word, *satyrion* (two-horned), refers to the two spurs of lip.

Distribution: Tropical and South Africa, W. Indian Ocean, Indian Subcontinent to SC. China.

Satyrium nepalense D. Don, Prodr. Fl. Nepal. 26. 1825; Hook.f., Fl. Brit. India 6: 168. 1890; C.E.C. Fisch., Fl. Madras 1476. 1928; Ramakrishna in Pullaiah, Fl. Andhra Pr. 3: 957. 1997; Misra *et al.,* Bull. Bot. Surv. India 50: 143. 2008; Seidenfaden in K.M. Matthew, Fl. Tamilnadu Carnatic 3: 1568. 1983.

Types: Nepal: Gossingsthan (Gofsain Than), *Wallich s.n.* (P, BM, GOET, MPU).

Etymology: This species was named after the type collection country, Nepal.

Terrestrial herbs, *ca* 50 cm high. Tubers 1-2, ovoid-oblong or elliptic-oblong, *ca* 4 × 2 cm. Leaves mostly basal and gradually passing into bracts, fleshy, sessile, ovate-lanceolate or oblong to linear-oblong or narrowly elliptic, acute, *ca* 10 × 5 cm. Inflorescence *ca* 20 cm long; spike cylindrical, densely many-flowered. Floral bracts longer than the ovary, greenish-pink, deflexed, ovate-lanceolate, obtuse, 3-veined, *ca* 1.6 × 0.6 cm. Pedicel with ovary greenish-pink, *ca* 6 mm long. Flowers non-resupinate, pink or rarely white, *ca* 1.2 cm across. Dorsal sepal oblanceolate-spathulate, obtuse, 3-veined, *ca* 6 × 1.5 mm; lateral sepals spreading, falcate, ovate or oblong-elliptic, obtuse, *ca* 5.5 × 2.6 mm. Petals twisted at the mouth of the lip, narrowly lanceolate or oblong-elliptic, obtuse, 3-veined, *ca* 5.5 × 1.2 mm. Lip hooded, globose or orbicular-oblong, 1 × 1.2 mm, reflexed rim, strongly keeled, 2-spurred; spurs parallel, close and longer than the ovary, cylindric, mouth funnel-shaped, *ca* 1.2 cm long. Column enclosed by lip, white, terete, *ca* 4 mm long, dilated at apex. Stigma 3-lobed; lobes flat, triangular; rostellum midlobe prominent, narrowly triangular. Anther pendent; anther locules parallel; pollinia 2, clavate, 0.8 mm. Capsules fusiform, *ca* 5 mm long.

Fl. & Fr.: September-December.

Habitat: Rare, it grows in marshy area in high altitude open grass lands or evergreen forest (800-1400 m).

Distribution: ***World*:** Bhutan, China, India, Myanmar, Nepal, Pakistan and Sri Lanka. ***India:*** Andhra Pradesh, Arunachal Pradesh, Jammu & Kashmir, Himachal Pradesh, Kerala, Manipur, Meghalaya, Nagaland, Sikkim and West Bengal. ***Eastern Ghats:*** *AP:* Visakhapatnam forest; *TN:* Shevaroy hills.

Specimens Examined: *AP:* Visakhapatnam forest, *VNS* 548 (CAL). *TN:* Shevaroy hills (SLM), *G. Bidie s.n.* (MH).

Schoenorchis Blume

Bijdr. Fl. Ned. Ind., 8: 361. 1825.

World:24/India:6/EG:1

Type: *Schoenorchis juncifolia* Blume

Etymology: The generic name is derived from the Greek words, *schoeno* (reed or rush) and *orchis* (orchid), refering to the rush-like leaves.

Distribution: India, Sri Lanka to Southern China and the Philippines, Thailand, south to Indonesia and east to New Guinea, Australia and the Pacific Islands.

Schoenorchis jerdoniana (Wight) Garay, Bot. Mus. Leafl. 23: 202. 1972; Seidenfaden in K.M. Matthew, Fl. Tamilnadu Carnatic 3: 1601. 1983. *Taeniophyllum jerdonianum* Wight, Icon. Pl. Ind. Orient. 5: t. 1756. 1851.

Type: India: Malabar, *Jerdon s.n.* (BLAT?).

Etymology: This species was named honour to T.C. Jerdon (1811-1872), who collected type material and birds and plants from India.

Epiphytic small herbs; stems tufted, *ca* 10 cm long. Leaves flat, linear, acute, *ca* 4 × 0.5 cm. Racemes longer than the leaves, simple or branched. Floral bracts longer than the ovary, ovate, acute. Flowers white. Sepals broadly ovate, 1-veined. Petals spreading, linear-lanceolate, 1-veined. Lip concave, oblong, 3-lobed; laterals rounded; midlobe oblong, undulate; spur saccate or sub-globose. Pollinia 4.

Fl.: March.

Habitat: Extremely rare, it grows on moss covered tree trunks (about 1200 m).

Distribution: Endemic to Kerala and Tamil Nadu. ***Eastern Ghats:*** *TN:* Kolli hills.

Specimens Examined: *TN:* Kolli hills (NK), *KMM* 25382 (RHT).

Seidenfadeniella C.S. Kumar

Cat. Ind. Orchids 43. 1994.

World:2/India:2/EG:1

Type: *Seidenfadeniella rosea* (Wight) C.S. Kumar

Etymology: The genus was named after well-known orchidologist, Gunnar Seidenfaden (1908-2001).

Distribution: South India and Sri Lanka.

Seidenfadeniella filiformis (Rchb.f.) Christenson & Ormerod, Fl. Palni Hills 3: 1258. 1999. *Schoenorchis filiformis* (Rchb.f.) Schltr., Repert. Spec. Nov. Regni Veg. Beih. 1: 986. 1913. *Saccolabium filiforme* Rchb.f., Ann. Bot. Syst. 6: 887. 1864. *Seidenfadeniella chrysantha* (Alston) C.S. Kumar, Cat. Ind. Orchids 47. 1994. *Saccolabium chrysanthum* Alston, Handb. Fl. Ceylon 6(Suppl.): 277. 1931.

Type: India: Anamally forest, *Cotton s.n.* (K).

Epiphtic herbs, *ca* 60 cm high. Stem terete. Leaves sessile, terete, *ca* 15 cm long. Inflorescence extra axillary, *ca* 6 cm long; racemes densely flowered. Floral bracts shorter than the ovary, ovate-lanceolate, acuminate, *ca* 2 mm long. Pedicel with ovary *ca* 6 mm long. Flowers orange-yellow and streaked with crimson brown lines. Dorsal sepal oblong, obtuse, *ca* 4.5 × 2 mm, 1-veined; lateral sepals spathulate or narrowly obovate, obtuse. Petals sub-orbicular, *ca* 4 mm long, 3-4-veined. Lip 3-lobed; lateral erect, orbicular; midlobe deflexed, ovate, papillate on

Spiranthes Rich.

De Orchid. Eur. 20. 28. 36. 1817 (*nom. cons.*).

World:42/India:2/EG:1

Type: *Spiranthes spiralis* (L.) Chevall. (lectotype)

Etymology: The generic name is derived from the Greek words, *speira* (spiral) and *anthes* (flowers) with reference to spirally arranged flowers of the species.

Distribution: Eurasia to SW Pacific, North Africa, North and Central America to Caribbean.

Spiranthes sinensis (Pers.) Ames, Orchidaceae 2: 53. 1908; C.E.C. Fisch., Fl. Madras 1454. 1928; Misra, Orchids Orissa 282. 2004; Seidenfaden in K.M. Matthew, Fl. Tamilnadu Carnatic 3: 1558. 1983. *Neottia sinensis* Pers., Syn. Pl. 2: 511. 1807. *Spiranthes australis* (R. Br.) Lindl., Bot. Reg. 10: t. 823. 1824; Hook.f., Fl. Brit. India 6: 102. 1890; Haines, Bot. Bihar & Orissa, 1161. 1924.

Type: 'Prope Cantonem Sinarum' *Loureiro s.n.* (holo:P-LOUR).

Etymology: This species was named after the type locality (Old name: China).

Terrestrial herbs, *ca* 45 cm high. Roots fasciculate. Stem ribbed, hairy. Leaves basal rosette, fleshy, sessile, linear-lanceolate or linear-oblong or oblanceolate, acuminate, 3-veined, *ca* 10 × 1.2 cm. Inflorescence spirally twisted, *ca* 30 cm long, secund; spike densely many-flowered, glandular hairy. Floral bracts as long as or longer than the ovary, pale green, ovate-lanceolate, acuminate, glandular hairy, *ca* 8 × 3 mm. Pedicel with ovary green, *ca* 6 mm long, glandular hairy. Flowers fragrant, white or whitish-pink, *ca* 4.5 mm across; outer surface of sepals and petals glandular hairy. Sepal oblong-lanceolate, acute or rounded, 1-veined, *ca* 4 × 1.6 mm. Petals liner-oblong or oblong-spathulate, rounded, 1-veined, *ca* 4 × 1 mm. Lip concave, 5-veined, *ca* 5.5 × 3 mm, obscurely 3-lobed; hypochile with two lateral oblong-ellipsoid white calli; laterals narrow; midlobe deflexed, oblong-obovate or subquadrate, truncate, margin crisped to irregularly crenulate, papillose within. Column white, obovate-oblong, *ca* 3 × 1.5 mm. Pollinia 4, obpyriform, 0.8 × 0.3 mm. Capsules fusiform, *ca* 6 mm long, with glandular hairy.

Fl. & Fr.: January-August.

Habitat: Occasional, it grows in marshy areas of open grass lands (above 200 m).

Distribution: ***World:*** Afghanistan, Australia, Bhutan, Borneo, Cambodia, China, India, Japan, Java, Korea, Malaysia, Manchuria, Mangolia, Moluccas, Myanmar, New Caledonia, New Guinea, New Zealand, Nepal, Pacific Is., Pakistan, Philippines, Samoa, Siberia, Solomon Is., Sri Lanka, Sulawesi, Sumatra, Taiwan, Tasmania, Thailand, Tibet, Timor, Russia, Vanuatu and Vietnam. ***India:*** Andhra Pradesh, Arunachal Pradesh, Assam, Bihar, Himachal Pradesh, Jammu & Kashmir, Karnataka, Kashmir, Kerala, Madhya Pradesh, Manipur, Meghalaya, Mizoram, Nagaland, Odisha, Punjab, Sikkim, Tamil Nadu, Tripura and West Bengal. ***Eastern Ghats:*** *AP:* Galikonda hills; *OD:* Similipal forest; *TN:* Shevaroy hills.

Specimens Examined : *AP:* Galikonda (VSKP), *GVS* 22658 (MH). *OD:* Similipal forest-Jenabila (MB), *Misra* 1058, 1059, 1060, 1061 (CAL). *TN:* Shevaroy hills (SLM), *AVNR* 27484 (MH), *DIA* 7801, *NV & TSJ* 22376 & 22389, *KMM* 27690 (RHT).

Staurochilus Ridl. *ex* Pfitzer

Nat. Pflanzenfam. Nachtr. (Engler & Prantl) 2: 16. 1900.

World:14/India:1/EG:1

Type: *Staurochilus fasciatus* (Rchb. f.) Ridl.

Etymology: The generic name is derived from the Greek words, *stauros* (cross) and *cheilos* (lip), refers to the parallel lateral lobes of the lip (cross).

Distribution: India, through SE Asia to Philippines and Indonesia.

Staurochilus ramosus (Lindl.) Seidenf., Opera Bot. 95: 95. 1988; Misra, Orchids Orissa 596. 2004; Misra *et al.,* Bull. Bot. Surv. India 50: 143. 2008. *Saccolabium ramosum* Lindl., Gen. Sp. Orchid. Pl. 224. 1833. *Cleisostoma ramosum* (Lindl.) Hook.f., Fl. Brit. India 6: 72. 1890.

Type: India: *Wallich* icon. (K (image!)).

Etymology: The specific epithet refers to the branched inflorescence of the species.

Epiphytic herbs; stem erect. Leaves lorate or oblong-lorate, *ca* 16 cm long. Panicles laxly many-flowered, flexuous. Floral bracts triangular-ovate. Pedicel with ovary greenish-brown. Flowers *ca* 6 mm across. Sepals and petals yellow-ochre with reddish or pink markings, oblong-obovate or oblong-oblanceolate. Lip white with maroon blotches, panduriform, 3-lobed; laterals quadrate or triangular; midlobe ovate, obscurely 3-lobed; disk papillose; spur cylindrical, pubescent, with tongue-shaped growth on back wall. Capsules cylindrical, *ca* 6 mm long.

Fl. & Fr.: April-September.

Habitat: Rarely grows in semi evergreen and moist deciduous forests (above 100m).

Distribution: ***World:*** Bangladesh, India, Malaysia, Myanmar and Thailand. ***India:*** Andhra Pradesh, Chhattisgarh, Jharkhand, Meghalaya, Odisha, Sikkim and West Bengal. ***Eastern Ghats:*** *AP:* Maredumilli forest; *OD:* Govindapali-Gupteswara forest, Kapilas hills, Rebana forest and Similipal forest.

Specimens Examined: *AP:* Maredumilli-Tiger point (EG), *SM* 0092 (RPRC). *OD:* Similipal forest-Kairakacha (MB), *SM* 1133 (CAL).

Taeniophyllum Blume

Bijdr. Fl. Ned. Ind. 8: 355, t. 70. 1825.

(World: 234/India:10/EG:1)

Type: *Taeniophyllum obtusum* Blume

Etymology: The generic name is derived from the Greek words, *taenio* (tapeworm) and *phyllon* (leaf), refering to the green photosynthetic roots of the genus.

Distribution: Ghana to Zimbabwe, Tropical and Subtropical Asia to Pacific.

Taeniophyllum alwisii Lindl., J. Proc. Linn. Soc., Bot. 3: 42. 1858; Hook.f., Fl. Brit. India 6: 76. 1890; Seidenfaden in K.M. Matthew, Fl. Tamilnadu Carnatic 3: 1609. 1983.

Iconotype: Sri Lanka: De Alwiz (icon).

Etymology: This species was named in honour of Haramanis de Alwis Seneviratne (1792-1894), botanical illustrator and also who illustrated type of the species from Sri Lanka.

Small epiphytic herbs, without stem and leaves. Roots green, trigonous or subterete. Inflorescence from the centre of the reduced stem, 1 cm long, 1-3-flowered, peduncle glabrous or glandular hairy. Floral bracts distichous, glandular, keeled, ovate, acute, 1.2 × 0.8 mm. Ovary sessile, 2 mm long. Flowers green or creamy white, *ca* 6 mm long. Sepals and petals united into tube, 1 mm long; sepal lobes oblong-lanceolate, obtuse or acute, 1 × 0.5 mm; petal lobes ovate, acute, 0.8 × 0.6 mm. Lip cymbiform, ovate, obtuse and recurved, 2.5 × 0.7 mm; spur oblong, *ca* 1 mm long. Stigmatic arms oblong. Anther unequally 4-locular; pollinia 4, pyriform, 0.2 mm. Capsules oblong, *ca* 1 cm long.

Fl. & Fr.: September-October.

Habitat: Extremely rare, it grows on densely moss-covered tree trunks of evergreen forests (above 1200 m).

Distribution: ***World:*** India and Sri Lanka. ***India:*** Karnataka, Kerala and Tamil Nadu. ***Eastern Ghats:*** *TN:* Shevaroy hills.

Specimens Examined: *TN:* Shavaroy hills (SLM), *KMM* & *DIA* 4981, 28569 (RHT).

Tainia Blume

Bijdr. Fl. Ned. Ind. 7: 354. 1825.

World:30/India:6/EG:1

Type: *Tainia speciosa* Blume

Etymology: The generic name is derived from Greek, refers either to the long petioles and leaves.

Distribution: Tropical and Subtropical Asia to NE Australia.

Tainia penangiana Hook.f., Fl. Brit. India 5: 820. 1890. *T. hookeriana* King & Pantl., J. Asiat. Soc. Bengal, Pt. 2, Nat. Hist. 64(2): 336. 1895; Misra, Orchids Orissa 496. 2004.

Syntypes: Malaysia: Penang Is., *Maingay s.n.* K & L (images!) & 1642 (CAL!).

Etymology: This species was named after the type locality area, Penang Is., Malaysia.

Terrestrial herbs; rhizome subterranean; pseudobulbs ovoid. Leaf oblong-lanceolate or linear-elliptic, petiole *ca* 30 cm long. Racemes laxly few-flowered. Floral bracts linear-lanceolate. Pedicel with ovary *ca* 6 mm long. Sepals and petals greenish-yellow with brown veins, elliptic-lanceolate, acuminate. Lip white flushed with purple dots, clawed, 3-lobed; laterals oblong; midlobe ovate or orbicular; disc with 3 yellow ridges; spur conical. Capsules oblong, *ca* 8 mm long.

Fl. & Fr.: January-March.

Habitat: Extremely rare, it grows along the perennial streams under shady conditions in humus-rich soils of semi-evergreen and moist deciduous forests (700-1000 m).

Distribution: ***World:*** China, India, Malaysia, Taiwan, Thailand and Vietnam. ***India:*** Arunachal Pradesh, Assam, Odisha, Sikkim and West Bengal. ***Eastern Ghats:*** *OD:* Similipal forest.

Specimens Examined: *OD:* Similipal forest-Jenabila (MB), *SM* 1064 & 1080 (CAL).

Taprobanea Christenson

Lindleyana 7: 90. 1992.

Monotypic genus

Type: *Taprobanea spathulata* (L.) Christenson

Etymology: The genus was named after the type locality country, Taprobanea (Ancient name of Sri Lanka).

Distribution: South India and Sri Lanka.

Taprobanea spathulata (L.) Christenson, Lindleyana 7: 91. 1992; Misra *et al.*, Bull. Bot. Surv. India 50: 143. 2008. *Vanda spathulata* (L.) Spreng., Syst. Veg. 3: 719. 1826; Hook.f., Fl. Brit. India 6: 50. 1890; C.E.C. Fisch., Fl. Madras 1444. 1928; Seidenfaden in K.M. Matthew, Fl. Tamilnadu Carnatic 3: 1599. 1983; Ramakrishna in Pullaiah, Fl. Andhra Pr. 3: 958. 1997. *Epidendrum spathulatum* L., Sp. Pl. 952. 1753.

Type: India: Malabar, Rheede, Hort. Malab. (Ponnampou Maravara) 12: t. 3. 1693.

Etymology: The specific epithet refers to the shape of petals of the species.

Epiphytic or lithophytic herbs, *ca* 1 m high. Stem scandent, erect, stout, nodes with roots. Leaves coriaceous, sessile, linear-oblong or linear-lanceolate, unequally bilobed, *ca* 9 × 2 cm, green spotted reddish-pink. Inflorescence axillary, much longer than the leaves, simple or branched, *ca* 80 cm long; raceme laxly many-flowered. Floral bracts greenish-yellow, fleshy, broadly ovate, acuminate, 3-veined, *ca* 6 × 4 mm. Pedicels with ovary greenish-yellow, *ca* 4 cm long. Flowers golden yellow, *ca* 3 cm across. Sepals spathulate-oblong or elliptic to oblong-ovate, obtuse, 7-veined, *ca* 1.6 × 1.1 cm. Petals obovate or spathulate, rounded, 7-veined, *ca* 1.5 × 0.9 cm. Lip sessile, *ca* 1.6 cm long, 3-lobed; laterals broadly triangular or subquadrate; midlobe flabellate, reniform, truncate, *ca* 1.1 × 1.5 cm; spur funnel-shaped, obtuse, *ca* 2 cm long. Column yellow, cylindrical, *ca* 6 × 3 mm; foot 2 mm long. Stigma sub-quadrate. Pollinia 2, globose-obovoid, 2 × 2 mm. Capsules fusiform, *ca* 6 cm long.

Fl. & Fr.: April-September.

Habitat: Rare, it grows under dense shady conditions of moist deciduous forests (above 700 m).

Distribution: ***World*:** India and Sri Lanka. ***India:*** Andhra Pradesh, Kerala and Tamil Nadu. ***Eastern Ghats:*** *AP:* Seshachalam hills; *TN:* Javvadhu hills, Kolli hills and Sirumalai hills.

Specimens Examined : *AP:* Mogalipenta (KDP), *BR & MVS* 34629 (SKU); Tirummala (CTR), *MVS & V.S.Rao* 32041 (SKU); Tirumala to Srivarimettu (CTR), *MVS & V.S.Rao* 31978 (SKU). Papanasanam valley (CTR), *MVS & KP* 340606 (SKU); Papanasanam dam site (CTR), *SM* 299 (RPRC). *TN:* Javvadhu hills (TVM), *KRK* 6066 (FRLH); Salem, *E.K. Krishanan s.n.* (DD); Kolli hills (NK), *KMM & CM* 1855 (RHT), *CR* 2064, *KMM & VAA* 3080, *DIA* 7610, *AM* 12991, *NV* 15897 (RHT); Sirumalai (DGL), *J. Pallithanam* 4460 (RHT); Pachamalai (TRP), *KMM* 4300 (RHT); Sengaattupatty, Pachamalai (TRP), *KMM* 2290, 2419, 4301 & 11595 (RHT); Kalrayan (SLM), *KMM & NV* 14006 (RHT); Kalrayan hills (SLM), *KMM* 62770 (RHT); Chitteri (SLM), *KMM & GP* 24298 (RHT).

Thunia Rchb.f.

Bot. Zeitung (Berlin) 10: 764. 1852.

World:6/India:2/EG:1

Type: *Thunia alba* (Lindl.) Rchb.f.

Etymology: The genus was named in honour of Count France A. Garf von Thun Hohenstein (1786-1873), a great orchid collector.

Distribution: Indian Subcontinent to SC. China and Peninsular Malaysia.

Thunia alba (Lindl.) Rchb.f. var. **bracteata** (Roxb.) N. Pearce & P.J. Cribb, Edinburgh J. Bot. 58: 116. 2001. *Limodorum bracteatum* Roxb., Fl. Ind. ed. 3: 466. 1832. *Thunia bracteata* (Roxb.) Schltr., Repert. Spec. Nov. Regni

Veg. Beih. 4: 205. 1919; Misra, Orchids Orissa 517. 2004. *T. venosa* Rolfe, Orchid Rev. 12: 206. 1905; Haines, Bot. Bihar & Orissa, 1167. 1924; Mooney, Suppl. Bot. Bihar & Orissa 210. 1956.

Type: India: Indiae Orientalis, *Smith s.n.* (not located).

Etymology: The specific epithet refers to the bracts nature.

Epiphytic or lithophytic herbs, *ca* 1 m high. Stem erect or pendulous, stout. Leaves membranous, sessile, elliptic-lanceolate or oblong-lanceolate, acuminate, *ca* 25 × 3.5 cm. Racemes compact, laxly 4-8-flowered. Floral bracts longer than the ovary, white, cymbiform, oblong-ovate, acute, *ca* 5 × 1.5 cm. Pedicel and ovary white, *ca* 2 cm long. Sepals and petals white, narrowly oblong-ovate or oblong, acute or acuminate, *ca* 6 × 1.2 cm. Lip white with purple veins, rhomboid or broadly oblong-pandurate, basal half convolute, margins crisped and irregularly denticulate, *ca* 5 × 4 cm; disc with 5-9 dentate–fimbriate ridges; spur cylindrical, *ca* 1.5 cm long. Column white, oblong, winged at apex, with foot. Pollinia 8, clavatc, 2 × 0.4 mm. Fruit cllipsoid, *ca* 3.5 cm long.

Fl. & Fr.: August-May.

Habitat: Rare, it grows on moss covered tree trunks and rocks of semi-evergreen and moist deciduous forests (700-1000 m).

Distribution: ***World:*** Bhutan, China, India, Myanmar, Nepal and Thailand. ***India:*** Andaman & Nicobar Is., Andhra Pradesh, Assam, Chhattisgarh, Jharkhand, Karnataka, Kerala, Maharashtra, Meghalaya, Odisha, Sikkim, Uttarakhand and West Bengal. ***Eastern Ghats:*** *AP:* Sapparla forest; *OD:* Rebana forest and Similipal forest.

Specimens Examined: *AP:* Sapparla RF (VSKP), *K.N. Reddy* 6693 (Herbarium, Laila Impex R & D Centre). *OD:* Similipal forest-Hatikhunti (MB), *SM* 713 (CAL); Lataput (KP), *SM* 8 (CAL).

Tropidia Lindl.

Edwards's Bot. Reg. 19: t. 1618. 1833.

World:35/India:5/EG:3

Type: *Tropidia curculigoides* Lindl.

Etymology: The generic name is derived from the Greek word, *tropideion* (keel), refers to the keeled lip of many species.

Distribution: Tropical and Subtropical Asia to SW Pacific, Florida to Tropical America.

Key to the species

1a. Spike terminal or axillary; lip with two large intra-marginal ridges...... **T. pedunculata**

1b. Raceme terminal; lip not as above 2

2a. Flowers ca 11 mm long; lip with yellow patch at the apex, hypochile without lamellae.. **T. angulosa**

2b. Flowers ca 5 mm long; lip without yellow patch, hypochile with pair of erect incurved lamellae.. **T. hegderaoi**

Tropidia angulosa (Lindl.) Blume, Coll. Orchid. 122. 1859; Hook.f., Fl. Brit. India 6: 92. 1890; C.E.C. Fisch., Fl. Madras 1452. 1928; Mooney, Suppl. Bot. Bihar & Orissa 208. 1956; Misra, Orchids Orissa 274. 2004; Ramakrishna in Pullaiah, Fl. Andhra Pr. 3: 958. 1997; Misra *et al.,* Bull. Bot. Surv. India 50: 143. 2008; Seidenfaden in K.M. Matthew, Fl. Tamilnadu Carnatic 3: 1556. 1983. *Cnemidia angulosa* Lindl., Edwards's Bot. Reg. 19: t. 1618. 1833.

Types: Bangladesh: Sylhet, Wall. Cat. no. 7388 (K).

Etymology: The specific epithet refers to the angular nature of rachis of the species.

Terrestrail herbs, *ca* 30 cm high. Roots fasciculate; rhizome subterannean, compact, woody; stem rigid, rarely branched. Leaves 2 at apex of the stem, plicate, opposite, sessile, elliptic-ovate or elliptic, margins undulate, apex acuminate, *ca* 12 × 6 cm. Inflorescence *ca* 8 cm long, angular; raceme densely many-flowered. Floral bracts longer than the ovary, pale green, linear, acuminate, *ca* 1 × 0.2 cm. Pedicel with ovary green, *ca* 1 cm long. Flowers non-resupinate, white. Sepals oblong-lanceolate, acute, 3-veined, *ca* 1.1 × 0.3 cm; laterals completely connate, apex 2-lobed, embracing lip. Petals keeled, ovate-lanceolate, acute, 3-veiend, *ca* 1 × 0.3 cm. Lip with yellow patch at the apex, cymbiform, oblong-oblanceolate or elliptic-oblong, obtuse, 3-veined, *ca* 1.1 × 4.5 cm; spur cylindrical, *ca* 4 mm long. Column white, cylindrical, 3 × 1 mm. Stigma triangular-ovate; rostellum long-lanceolate, bifid. Pollinia 2, clavate, 2 × 0.6 mm. Capsules ellipsoid-cylindric, *ca* 2 cm long.

Fl. & Fr.: August-November.

Habitat: Occasional, it grows under dense shady conditions in humus rich soils along the near streams of semi evergreen forests (700-1100 m).

Distribution: ***World:*** Bangladesh, Bhutan, Borneo, China, India, Java, Laos, Malaysia, Myanmar, Sumatra, Taiwan, Thailand and Vietnam. ***India:*** Andhra Pradesh, Arunachal Pradesh, Assam, Chhattisgarh, Karnataka, Kerala, Meghalaya, Odisha, Sikkim, Tamil Nadu, Tripura and West Bengal. ***Eastern Ghats:*** *AP:* Maredumilli forest; *OD:* Similipal forest; *TN*: Kolli hills .

Specimens Examined: *AP:* Maredumilli (EG), *BR* & *KNR* 10603 (SKU); Lankamalai hills (KDP), *SRSR* 14636 (SKU); Dummakonda (EG), *GVS* 68629 (MH & CAL) & *SS* 3665 (AU). *OD:* Similipal forest-Jenabila (MB), *SM* 1705 & 1632 (CAL); Similipal forest-UBK (MB), *SM* 1819 (CAL). *TN*: Kolli hills (NK), *PP* 21706, *KM* & *GP* 24122, *KMM* 26684, *SJB* 27937 (RHT); Melagiris (DMP), *KMM* 63466 (RHT).

Tropidia hegderaoi S. Misra, Nelumbo 54: 13. 2012.

Type: India: Tamil Nadu, Kolli hills (Kollimalais), Ariyur shola, Mathikettan, *SM* 2525 (holo : CAL).

Etymology: This species was named in honour of two Indian orchidologists, Dr. Sadananda Hegde and Dr. A. Nageswar Rao.

Terrestrial herbs, *ca* 24 cm high; stem woody. Leaves 2, at the apex of the stem, plicate, sessile, elliptic, acuminate, *ca* 13 × 4 cm. Inflorescence terminal, *ca* 3.5 cm long; raceme densely 8-10-flowered. Floral bracts green, subulate, acute, *ca* 3.25 × 1.25 mm. Pedicel and ovary greenish-white, *ca* 8 mm long. Flowers non-resupinate, white, *ca* 5 mm long. Dorsal sepal elliptic-lanceolate, 3-veined, *ca* 4.2 × 2.5 mm; lateral sepals connate along inner margins near the apex. Petals elliptic-ovate or obovate-lanceolate, acute or acuminate, 3-veined, *ca* 4 × 2.8 mm. Lip cymbiform, broadly obovate to sub-orbicular, *ca* 3 × 3 mm; hypochile with a pair of lamellae. Stigma entire. Pollinia 2, clavate, 1.4 × 0.6 mm. Capsules oblong-ellipsoid, *ca* 1 cm long.

Fl. & Fr.: July-August.

Habitat: Extremely rare, it grows under dense shady condition in humus rich soils of evergreen and moist deciduous forests (1100-1450 m).

Distribution: Endemic to Kolli hills (Tamil Nadu).

Specimens Examined: *TN:* Kolli hills (Kollimalais), Ariyur shola, Mathikettan, *SM* 2526, 2527, 2528 (OSC).

Tropidia pedunculata Blume, Coll. Orchid. 122. 1859; Misra, Orchids Orissa 277. 2004. *T. maingayi* Hook.f., Fl. Brit. India 6: 93. 1890.

Syntypes: Sumatra: Padang, *Korthals s.n.* (K & L); Palembang, *Praetorius s.n.* (BO).

Etymology: The specific epithet refers to long peduncle of the species.

Terrestrial herbs, *ca* 45 cm high; rhizome sub-terranean, woody; stem branched or unbranched. Leaves elliptic-lanceolate, acuminate. Inflorescence terminal or axillary; spike sub-capitate, densely many-flowered. Floral bracts as long as or longer than the ovary, subulate. Pedicel with ovary *ca* 5 mm long. Flowers greenish-white, *ca* 6.5 mm long. Sepals oblong-lanceolate or oblong-ovate; petals oblong-lanceolate. Lip oblong, reflexed, with two intra-marginal ridges. Capsules ellipsoid.

Fl. & Fr.: June-December.

Habitat: Extremely rare, it grows near the streams under deep shady condition in humus rich soils of moist deciduous forests (above 400 m).

Distribution: ***World:*** Bangladesh, India, Indonesia, Malaysia, Myanmar, Philippines and Thailand. ***India:*** Arunachal Pradesh, Kerala, Manipur, Meghalaya, Nagaland, Odisha, Tamil Nadu, Uttarakhand and West Bengal. ***Eastern Ghats:*** *OD:* Barbara-Dhuanali forest, Rebana forest and Similipal forest.

Specimens Examined: *OD:* Purunapani nala (BLW), *SM* 1771 & 1772 (CAL).

Vanda R. Br.

Bot. Reg. 6: t. 506. 1820.

World:74/India:11/EG:2

Type: *Vanda tessellata* (Roxb.) Hook. *ex* G. Don

Etymology: The generic name is derived from Sanskrit word, *Vanda* meaning epiphyte.

Distribution: Tropical & Subtropical Asia to NW. Pacific.

Key to the species

1a. Flowers more than 4 cm across; sepals and petals brown tessellated; capsule more than 6 cm long.. **V. tessellata**

1b. Flowers less than 2 cm across; sepals and petals not tessellated; capsule less than 4 cm long... **V. testacea**

Vanda tessellata (Roxb.) Hook. *ex* G. Don, Hort. Brit. 372. 1830; C.E.C. Fisch., Fl. Madras 1445. 1928; Haines, Bot. Bihar & Orissa, 1181. 1924; Misra, Orchids Orissa 635. 2004; Seidenfaden in K.M. Matthew, Fl. Tamilnadu Carnatic 3: 1600. 1983; Ramakrishna in Pullaiah, Fl. Andhra Pr. 3: 959. 1997; Misra *et al.,* Bull. Bot. Surv. India 50: 143. 2008. *Epidendrum tessellatum* Roxb., Pl. Coromandel 1: 34. 1795. *V. roxburghii* R. Br., Bot. Reg. 6: t. 506. 1820; Hook.f., Fl. Brit. India 6: 52. 1890.

Types: India: Andhra Pradesh, Circar mountains, *Roxburgh s.n.* (CAL, BM).

Etymology: The specific epithet refers to the tessellated flower of the species.

Epiphytic herbs. Stem erect or pendulous, woody. Leaves coriaceous, sessile, linear-oblong, unequally bilobed, *ca* 25 × 3 cm. Inflorescence axillary, erect, *ca* 30 cm long; raceme laxly few-flowered. Floral bracts greenish-white, broadly triangular-ovate, obtuse, 1-veined, *ca* 4 × 3 mm. Pedicel with ovary pale white, *ca* 6 cm long. Flowers fragrant, grey or greyish-blue, buff or red and brown tessellated, margin white, *ca* 5 cm across. Sepals and petals spreading, clawed, coriaceous, twisted, obovate, obtuse, margins strongly waved, 5-7-veined; dorsal *ca* 2.5 × 1.3 cm; laterals *ca* 2.5 × 1.6 cm; petals *ca* 2.5 × 1.1 cm. Lip sessile, fleshy, *ca* 1.5 cm long, 3-lobed; laterals falcate, oblong-ovate; midlobe panduriform, truncately bifid apex, with wide sinus in between; disc with fleshy ridges; spur laterally compressed, conical, *ca* 5 mm long. Column white, cylindrical, *ca* 5 × 3 mm; foot indistinct. Stigma sub-orbicular. Pollinia 2, globular, 2.5 × 2.5 mm. Capsules oblong, *ca* 12 cm long.

Figure 8. **Vanda tessellata** (Roxb.) Hook. ex Don
A. Twig, B. Sepal, C. Petal, D. Lip front view, E. Operculum with pollinia, F. Pollinium

Fl. & Fr.: May-August.

Habitat: Common in dry deciduous forest (above 400 m).

Distribution: ***World:*** Bangladesh, Bhutan, Borneo, China, India, Java, Myanmar, Nepal and Sri Lanka. ***India:*** Andhra Pradesh, Bihar, Chhattisgarh, Gujarat, Jharkhand, Karnataka, Kerala, Maharashtra, Madhya Pradesh, Odisha, Punjab, Rajasthan, Tamil Nadu, Tripura, Uttarakhand, Uttar Pradesh and West Bengal. ***Eastern Ghats:*** Common in all parts of Eastern Ghats.

Specimens Examined: *AP:* Gundlbrahmeswaram (KNL), *BR & BSS* 30529 (SKU); Chelama (KNL), *KNR & DAM* 10634 and *JLE* 16781 (MH); Velgode RF (KNL), *SKW* 4837; Siddhaleru–Kangemadugu (CTR), *GVS Murthy* 118258 (BSID); Guvvalacheruvu (KDP), *R.V. Reddy* 8461 (SKU); Panchalingala valley (KDP) *SRSR* 14634 (SKU); Lankamalai hills (KDP), *SRSR* 14634 (SKU); Panapakam-Amudalakona (CTR), *GVS* 31853 (MH); Veligonda hills (NLR), *MSR* 1337 (CAL); Penchalakona (NLR), *ASR* 6413 (VV); Malleswarakona – Venkatagiri (NLR), *ASR* 6519 (BSID); Rampachodavaram (EG), *BR & KP* 34662 (SKU); Gokavaram (EG),*GVS* 24201 (MH); Lakshmipuram (VSKP), *CAB* 1794 (MH); between Kurupam and Levidi (SKLM), *GVS* 62483 (MH & CAL). *OD:* Similipal forest (MB), *D.K. Agrawala s.n.* (CAL); Similipal forest (MB), *Bahali & Agrawala* 494 (CAL); Rebana forest (KJ), *G. Panigrahi* 8470 (CAL); Deogam village (Dhenkenal), *sine coll.* 11282 (BSID). *TN:* Anchetty RF (DMP), *EV* 57984 (MH); Billigundala (SLM), *EV* 24115 (MH); Javvadhu hills (TVM), *RVS* 43614 (FRLH); Pachamalai (TRP), *DIA & CM* 9056, *A. Diraviam & M.S. Xavier* 20701, *KMM & NV* 15648 (RHT); Chinnakalrayan (SLM), *DIA* 8302, *NV* 14368 (RHT); Periyakalrayan (SLM), *NV & CM* 17642, *KMM* 23556 (RHT); Kolli hills, Solakkadu (NK), *KMM & VKN* 9671 (RHT).

Vanda testacea (Lindl.) Rchb.f., Gard. Chron. n.s., 8: 166. 1877; Misra, Orchids Orissa 631. 2004; Seidenfaden in K.M. Matthew, Fl. Tamilnadu Carnatic 3: 1599. 1983; Ramakrishna in Pullaiah, Fl. Andhra Pr. 3: 959. 1997; Misra *et al.,* Bull. Bot. Surv. India 50: 143. 2008. *V. parviflora* Lindl., Edwards's Bot. Reg. 30 (Misc.): 45. 1844; Hook.f., Fl. Brit. India 6: 50. 1890; C.E.C. Fisch., Fl. Madras 1444. 1928; Haines, Bot. Bihar & Orissa, 1181. 1924.

Type: Sri Lanka: *Macrae s.n.* (K?).

Etymology: The specific epithet refers to the brownish-yellow lip of the species.

Epiphytic herbs. Stem erect, stout. Leaves coriaceous, sessile, linear-oblong, unequally lobed, *ca* 20 × 2 cm. Inflorescence axillary, erect, *ca* 25 cm long; raceme laxly many-flowered. Floral bracts brownish-green, broadly triangular, acute, 3-veined, *ca* 2.2 mm. Pedicel with ovary yellowish-green, *ca* 2.5 cm long. Flowers yellow with brownish-pink or white lip, *ca* 1.8 cm across. Dorsal sepal spathulate, rounded, 5-8-veined, *ca* 1.2 × 0.6 cm; lateral sepals twisted, broadly obovate, rounded, *ca* 1.2 × 0.8 cm. Petals twisted, obovate-spathulate, rounded, 5-veined, *ca* 1.2 × 0.5 cm. Lip sessile, 3-lobed; laterals triangular, *ca* 3 mm long; midlobe reflexed, oblong-subpanduriform, margins crenulated, apex retuse, *ca* 8 × 4.2 mm; disk with two fleshy, tubercled, purple-ridges and three short fleshy bands; spur laterally compressed, cylindrical, *ca* 4 mm long. Column yellow with purple dotted, cylindrical-truncated, *ca* 3 × 2 mm, with distinct foot. Stigma sub-quadrate. Pollinia 2, sub-globose, 1.8 × 1.8 mm. Capsules oblong-elliptic, *ca* 4 cm long.

Fl. & Fr.: March-August.

Habitat: Occasional, it grows in dry and moist deciduous forests (above 500 m).

Distribution: ***World:*** Bhutan, China, India, Myanmar, Nepal, Sri Lanka and Thailand. ***India:*** Andhra Pradesh, Arunachal Pradesh, Assam, Bihar, Chhattisgarh, Himachal Pradesh, Gujarat, Jharkhand, Karnataka, Kerala, Madhya

Pradesh, Maharashtra, Mizoram, Nagaland, Odisha, Rajasthan, Sikkim, Tamil Nadu and Uttarakhand. ***Eastern Ghats:*** Common in Eastern Ghats.

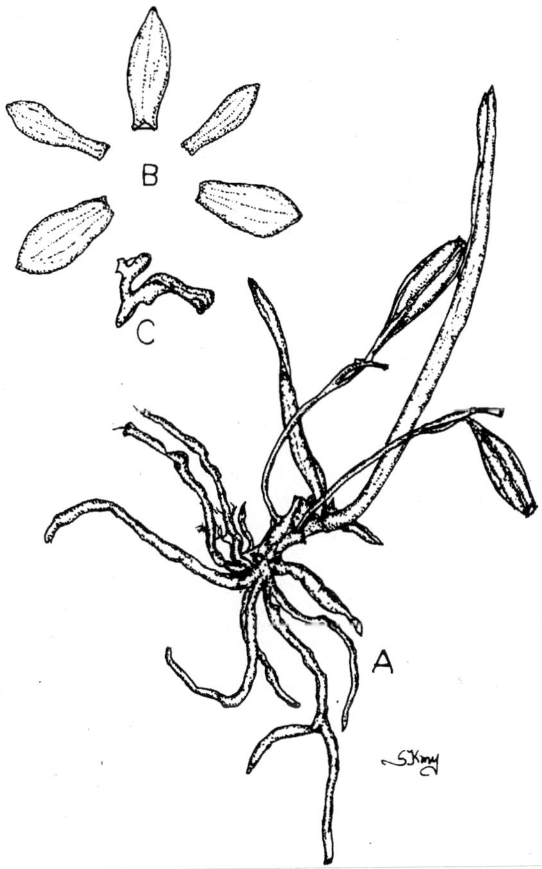

Figure 9. **Vanda testacea** (Lindl.) Rchb.f.
A. Habit; B. Sepals & petals; C. Lip

Specimens Examined: *AP:* Gundlabrameswaram (KNL), *RVR* 2386 (SKU); Kolanubarathi (KNL), *BSS* & *SKB* 34184 (SKU); Ramanapenta to GBM (KNL), *BR* & *KP* 34637 (SKU); Tirumala to Srivarimetlu (CTR), *MVS* & *V.S.Rao* 31989 (SKU); Bata – Atmakur (NLR), *ASR* 4729 (BSID); Somasila (NLR), *ASR* 6456; Venkatagiri (NLR), *ASR* 6390 & 8727 (VV); Rampa hills (EG), *KS* 23455 (SKU); Rajavomangi (EG), *SS* 777 (AU); Ettukkuru, Maredumilli (EG), *MM* 105130 (MH); Bhupathipaliam (EG), *VNS* 266 (CAL); Sesharayi-Yarlagadda (EG), *VNS* 675 (CAL); Mudurlanka, Maredumilli (EG), *MM* 101252 (MH); Muzumamidivalasa – Maredumilli (EG), *MM* 101201 (BSID); Galikonda (VSKP), *GVS* 19658 (MH); Lakshmipuram (VSKP), *CAB* 1796 (MH); Minumaluru (VSKP), *GVS* 30025 (MH); Chintapalli (VSKP), *SM* 87 (RPRC). *OD:* Laxmipur (KP), *SM* 18 (CAL); Gummada (RG), *GVS* 62451 (MH

& CAL). *TE*: Kudichintalabailu (MBNR), *TDCK* 15385 (SKU); Rampur (MBNR), *BSS* & *KP* 34609 (SKU). *TN:* Melpat (SA), *CAB* 1027 (MH); Shevaroys hills (SLM), *AVNR* 23072 (MH); Pulluhalli (SLM), *CAB* 9897 (MH); Javadi hills, Kanimanthur (NA), *MBV* 1058 (MH); Javvadhu hills (TVM), *KRK* & *RVS* 71420, 73227 (FRLH); Sirumalai (DGL), *SKS* 2613 (SGH); Sirumalai hills (DGL), *KRK* & *RVS* 106222 (FRLH); Valasamalai RF, Thenmalai MPCA (TVM), *KRK* 6011 (FRLH).

Vanilla Plum. *ex* Mill.

Gard. Dict. Abr. ed. 4. s.p. 1754.

World:65/India:6/EG:2

Type: *Vanilla mexicana* Mill. (lectotype)

Etymology: The generic name derived from Spanish, *vaina* means small seeds of fruit.

Distribution: Tropical and Subtropical.

Key to the species

1a. Flowers *ca* 6.5 cm long; disc of lip with 2 parallel, fimbriate lamellate with short papillous area opposite the stigma........................ **V. walkeriae**

1b. Flowers *ca* 4.5 cm long; disk of lip with 2 fimbriate lamellae on proximal half of short area opposite the stigma with close flat tile shaped scaly hairs........................ **V. wightii**

Vanilla walkeriae Wight, Icon. Pl. Ind. Orient. 3: t. 932. 1845; Hook.f., Fl. Brit. India 6: 90. 1890; C.E.C. Fisch., Fl. Madras 1451. 1928; Misra *et al.,* Bull. Bot. Surv. India 50: 144. 2008; Seidenfaden in K.M. Matthew, Fl. Tamilnadu Carnatic 3: 1570. 1983.

Type: Sri Lanka: *Walker s.n.* (K (image!)).

Etymology: This species was named in honour of General George Warren Walker (1778-1844), who collected plants from India and Sri Lanka (Ceylon).

Terrestrial, less robust, climbing herbs, *ca* 6 m high. Roots branched; nodal roots clasping the tree trunks. Stem greenish with pink dots, fleshy, stout, sub-cylindrical, *ca* 1 mm thick, nodded, with 2 or 3 longitudinal grooves, branched. Leaves deciduous, fleshy, sessile, ovate-lanceolate, acuminate, *ca* 2 × 0.5 cm. Inflorescence axillary, *ca* 15 cm long; raceme congested, densely few-flowered. Floral bract light green, reflexed, broadly ovate, obtuse, 1-3-veined, *ca* 2 × 0.8 cm. Pedicel with ovary green, *ca* 5 cm long. Flowers sweet-scented, fleshy, white, *ca* 6.5 cm across. Sepals concave, oblong-lanceolate or oblanceolate, acute; dorsal sepal *ca* 5.5 × 1.5 cm; lateral sepals *ca* 5.3 × 1.7 cm. Petals spathulate-oblanceolate, acute, *ca* 5 × 2 cm. Lip clawed, trumpet-shaped, rhomboid when spread out, undulate, acute, *ca* 5 × 3 cm, obscurely 3-lobed; disc with 2 parallel, fimbriate lamellate with short papillous area opposite the stigma. Column white, elongated, obovate-oblong, *ca* 2 × 1.2 cm. Stigma obovate; rostellum curved downwards and covered stigma. Pollinia 2, ovate, 2.5 × 2 mm, without stipes. Capsules dark green, cylindrical, *ca* 15 cm long.

Fl. & Fr.: March-June.

Habitat: Extremely rare, it grows in open scrub and dry deciduous forests (above 300 m).

Distribution: ***World***: India and Sri Lanka. ***India:*** Andhra Pradesh, Kerala and Tamil Nadu. ***Eastern Ghats:*** AP: Addatigala forest.

Specimens Examined: *AP:* Addatigala (EG), *BR* & *KP* 41609 (SKU); *KS* 23238 (SKU); Near Lubarthi (EG), *SS* 7611 (AU); Surampalem, Rajavomangi reserve forest (EG), *R.S. Rao* 10193 (AU). *TN*: Kolli hills (NK), *PP* 23519, *KMM* 26089, *Diraviam & Anthony* 27780 (RHT); Chinnakalrayan (SLM), *KMM* 24772, 26999 & 63665, *KMM*, *SJB* & *NRI* 29160 (RHT); Pachamalai (TRP), *KMM* 63865 (RHT).

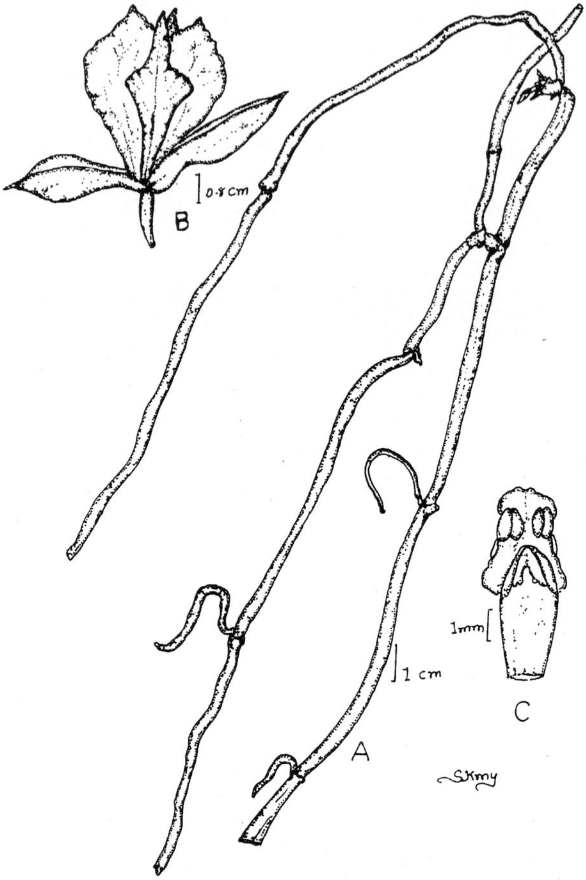

Figure 10. **Vanilla walkeriae** Wight
A. Habit; B. Flower; C. Lip and column

Vanilla wightii Lindl. *ex* Wight, Icon. Pl. Ind. Orient. 3: t. 931. 1845; Seidenfaden in K.M. Matthew, Fl. Tamilnadu Carnatic 3: 1570. 1983.

Types: India: Kerala, Travancore, Trivandrum, Wight 2091 (K (images!)).

Etymology: This species was named after the well-known taxonomist from India, R. Wight (1796-1872), who collected type material.

This species is much similar to above species but differs: plants more robust, *ca* 8 m high; flowers small, white, *ca* 3 cm long; sepals linear-oblong; petals lanceolate; lip 2-2.5 × 1-1.5 cm, with rounded lateral lobe and midlobe triangular; disk with 2 fimbriate lamellae on proximal half of short area opposite the stigma with close flat tile shaped scaly hairs.

Fl. & Fr.: March-April.

Habitat: Extremely rare, it grows in open scrub and dry deciduous forests (400-600 m).

Distribution: Endemic to India (Andhra Pradesh, Kerala and Tamil Nadu). ***Eastern Ghats:*** *TN:* Pacchaimalais.

Specimens Examined: *TN:* Pacchaimalais (TRP), *KMM* 27400, 29401 & 29402, *SJB* 29403 (RHT).

Zeuxine Lindl.

Coll. Bot., App. 1. 1826 (*nom. cons.*).

World:90/India:19/EG: 9

Type: *Zeuxine sulcata* (Roxb.) Lindl.

Etymology: The generic name is derived from Greek word, *zeuxis* (unite), refers to the partial union of the lip and column.

Distribution: Tropical Africa to Central Asia and W. Pacific.

Key to the species

1a. Epichile of lip with two divergent lobules 2
1b. Epichile of lip not as above 7
2a. Epichile lobules small, *ca* 3 mm long 3
2b. Epichile lobules large, more than 5 mm long 4
3a. Leaves with silvery-grey band along median veins; column with distinct wing .... **Z. nervosa**
3b. Leaves without band; column without wing .... **Z. gracilis**
4a. Flowers less than 1.5 cm long 5
4b. Flowers more than 1.5 cm long 6
5a. Epichile lobules longer and broader, 4-8 x 3-5 mm .... **Z. grandis**
5b. Epichile lobules relatively smaller, 2-3 x 2-2.5 mm .... **Z. affinis**
6a. Lateral sepals ovate-lanceolate; epichile lobules with entire margins .... **Z. chowdheryi**
6b. Lateral sepals ovate-oblong to awl-shaped; epichile lobules with serrated margins .... **Z. longilabris**
7a. Leaves ovate-lanceolate .... **Z. mooneyi**
7b. Leaves linear 8
8a. Labellum white and epichile 1-lobed .... **Z. lindleyana**
8b. Labellum yellow and epichile 2-lobed .... **Z. strateumatica**